Universitext

*Editorial Board
(North America):*

S. Axler
F.W. Gehring
P.R. Halmos

Springer
*New York
Berlin
Heidelberg
Barcelona
Budapest
Hong Kong
London
Milan
Paris
Santa Clara
Singapore
Tokyo*

Universitext

Editors (North America): S. Axler, F.W. Gehring, and P.R. Halmos

Aksoy/Khamsi: Nonstandard Methods in Fixed Point Theory
Aupetit: A Primer on Spectral Theory
Booss/Bleecker: Topology and Analysis
Borkar: Probability Theory: An Advanced Course
Carleson/Gamelin: Complex Dynamics
Cecil: Lie Sphere Geometry: With Applications to Submanifolds
Chae: Lebesgue Integration (2nd ed.)
Charlap: Bieberbach Groups and Flat Manifolds
Chern: Complex Manifolds Without Potential Theory
Cohn: A Classical Invitation to Algebraic Numbers and Class Fields
Curtis: Abstract Linear Algebra
Curtis: Matrix Groups
DiBenedetto: Degenerate Parabolic Equations
Dimca: Singularities and Topology of Hypersurfaces
Edwards: A Formal Background to Mathematics I a/b
Edwards: A Formal Background to Mathematics II a/b
Foulds: Graph Theory Applications
Fuhrmann: A Polynomial Approach to Linear Algebra
Gardiner: A First Course in Group Theory
Gårding/Tambour: Algebra for Computer Science
Goldblatt: Orthogonality and Spacetime Geometry
Hahn: Quadratic Algebras, Clifford Algebras, and Arithmetic Witt Groups
Holmgren: A First Course in Discrete Dynamical Systems (2nd ed.)
Howe/Tan: Non-Abelian Harmonic Analysis: Applications of $SL(2, R)$
Howes: Modern Analysis and Topology
Humi/Miller: Second Course in Ordinary Differential Equations
Hurwitz/Kritikos: Lectures on Number Theory
Jennings: Modern Geometry with Applications
Jones/Morris/Pearson: Abstract Algebra and Famous Impossibilities
Kannan/Krueger: Advanced Analysis
Kelly/Matthews: The Non-Euclidean Hyperbolic Plane
Kostrikin: Introduction to Algebra
Luecking/Rubel: Complex Analysis: A Functional Analysis Approach
MacLane/Moerdijk: Sheaves in Geometry and Logic
Marcus: Number Fields
McCarthy: Introduction to Arithmetical Functions
Meyer: Essential Mathematics for Applied Fields
Mines/Richman/Ruitenburg: A Course in Constructive Algebra
Moise: Introductory Problems Course in Analysis and Topology
Morris: Introduction to Game Theory
Porter/Woods: Extensions and Absolutes of Hausdorff Spaces
Ramsay/Richtmyer: Introduction to Hyperbolic Geometry
Reisel: Elementary Theory of Metric Spaces
Rickart: Natural Function Algebras
Rotman: Galois Theory

(continued after index)

Richard A. Holmgren

A First Course in Discrete Dynamical Systems

Second Edition

With 56 Figures

Springer

Richard A. Holmgren
Department of Mathematics
Allegheny College
Meadville, PA 16335-3902
USA

Editorial Board
(North America):

S. Axler
Department of Mathematics
Michigan State University
East Lansing, MI 48824
USA

F.W. Gehring
Department of Mathematics
University of Michigan
Ann Arbor, MI 48109
USA

P.R. Halmos
Department of Mathematics
Santa Clara University
Santa Clara, CA 95053
USA

Mathematics Subject Classification (1991): 39A12, 58F13, 58F20

Library of Congress Cataloging-in-Publication Data
Holmgren, Richard A.
  A first course in discrete dynamical systems/Richard A.
Holmgren. — 2nd ed.
    p.   cm. — (Universitext)
  Includes bibliographical references and index.
  ISBN 0-387-94780-9 (soft:alk. paper)
  1. Differentiable dynamical systems. I. Title.
QA614.8.H65 1996
514′.74—dc20   96-14777

Printed on acid-free paper.

© 1996 Springer-Verlag New York, Inc.
All rights reserved. This work may not be translated or copied in whole or in part without the written permission of the publisher (Springer-Verlag New York, Inc., 175 Fifth Avenue, New York, NY 10010, USA), except for brief excerpts in connection with reviews or scholarly analysis. Use in connection with any form of information storage and retrieval, electronic adaptation, computer software, or by similar or dissimilar methodology now known or hereafter developed is forbidden.
The use of general descriptive names, trade names, trademarks, etc., in this publication, even if the former are not especially identified, is not to be taken as a sign that such names, as understood by the Trade Marks and Merchandise Marks Act, may accordingly be used freely by anyone.

Production managed by Hal Henglein; manufacturing supervised by Joe Quatela.
Camera-ready copy prepared from the author's AMS-LaTeX files.
Printed and bound by R.R. Donnelley & Sons, Harrisonburg, VA.
Printed in the United States of America.

9 8 7 6 5 4 3 2 1

ISBN 0-387-94780-9 Springer-Verlag New York Berlin Heidelberg   SPIN 10538102

To Carol and our twins
David Sequoy and Ellen Topeah

Preface

An increasing number of colleges and universities are offering undergraduate courses in discrete dynamical systems. This growth is due in part to the proliferation of inexpensive and powerful computers, which have provided access to the interesting and complex phenomena that are at the heart of dynamics. A second reason for introducing dynamics into the undergraduate curriculum is that it serves as a bridge from concrete, often algorithmic, calculus courses to the more abstract concepts of analysis and topology.

Discrete dynamical systems are essentially iterated functions, and if there is one thing computers do well, it is iteration. It is now possible for anyone with access to a personal computer to generate beautiful images the roots of which lie in discrete dynamical systems. The mathematics behind the pictures is beautiful in its own right and is the subject of this text. Every effort has been made to exploit this opportunity to illustrate the beauty and power of mathematics in an interesting and engaging way. This work is first and foremost a mathematics book. Individuals who read it and do the exercises will gain not only an understanding of dynamical systems, but an increased understanding of the related areas in analysis as well.

Rationale for the new edition. After completing the first edition of this text, I thought that I had said what I wanted to say about dynamics and didn't expect to substantially revise my work. However, shortly after publishing the text, my students convinced me that there was no compelling reason to treat symbolic dynamics and metric spaces before introducing the concept of chaos. Further, one can study the dynamics of Newton's method and complex dynamics without ever studying symbolic dynamics or introducing metric spaces. Since metric spaces and symbolic dynamics, played a central role in the first edition beginning in Chapter 9, I set out to rewrite the core of the text. This edition is the result of my efforts.

The major changes are to introduce the notion of chaos for real functions in Chapter 8 and postpone the introduction of metric spaces and symbolic dynamics until the optional Chapter 11. These changes have necessitated the complete rewriting of Chapters 8, 9, 10, and 11. Additional changes include the rewriting of the proof of the special case of Sarkovskii's theorem in Chapter 5. I believe the new proof is much easier to follow. Some new exercises have been added, and many of the more difficult exercises have had hints added to make them more accessible to the typical undergraduate. Lemma 2.10 has been added in Chapter 2 and used in subsequent chapters to greatly simplify some of the proofs and exercises. Finally, the *Mathematica*® code in the appendix has been optimized.

How to use this book. This text is suitable for a one-semester course on discrete dynamical systems. It is based on notes from undergraduate courses that I have taught over the last few years. The material is intended for use by undergraduate students with a year or more of college calculus behind them. Students in my courses have come from numerous disciplines; most have been majors in other disciplines who are taking mathematics courses because they have a general interest in the subject. Concepts from calculus are reviewed as necessary. In particular, Chapters 2 and 3 are devoted to a review of functions and the properties of the real numbers. My students have found the material in these chapters to be extremely useful as background for the subsequent chapters. Other concepts are reviewed or introduced in later chapters.

The interdependence of Chapters 1 to 9 is fairly deep, and these should be covered sequentially. Students with a good background in real analysis can skip Chapters 2 and 3. On the other hand, students with only a year of calculus and little or no experience reading and writing mathematical proofs are *especially* encouraged to read these chapters and do the exercises. They are intended to provide the mathematical sophistication necessary to handle the remaining portions of the book. Readers interested in moving through the material quickly may wish to treat Chapters 5 and 7 lightly; only an understanding of the terminology is necessary for subsequent chapters. Chapters 10, 11, 12, 13, and 14 can be done in virtually any order, though there are a few interdependencies. In particular: the doubling map defined in Exercise 11.14 is used in Example 12.1 and Example 14.15; the topology of the complex plane is defined in terms of a metric space, so readers who skip Chapter 11 will need to fill in a small amount of background, though that isn't hard; Section 14.5 has Chapter 12 as a prerequisite; and the first three sections of Chapter 14 are prerequisites for Chapter 15.

Since the heart of any good mathematics textbook is the exercises, I have provided a liberal quantity of interesting ones. The exercises range from computational to those requiring a proof. A large number of the exercises involving the theory can, at the instructor's discretion, be answered with descriptive paragraphs or drawings rather than formal proofs. Some of the exercises are assumed later in the text. These are marked by a black dot (•) and should not be skipped. Particularly difficult exercises are marked by a star (∗) or, in some cases, a double star (∗∗).

All students are encouraged to tackle the star problems. Trying to solve them deepens one's understanding of the material, even if the particular exercise is never completed. Indeed, there are one or two for which I do not have a complete solution, but they are very interesting (and fun) to work on.

In some cases, it is nearly impossible to complete an exercise without assistance from a computer. *Mathematica* or a similar package is an excellent resource for doing most of them. The relevant *Mathematica* code is provided in the Appendix. Electronic versions of the code may be obtained by contacting the author directly. (The author's addresses are found on the copyright page and at the end of the Preface.) It is also very easy to write simple programs that will assist with the exercises. Details and sources of more information are provided in the references and the Appendix.

Acknowledgments. Numerous individuals have assisted in the development of this text. First and foremost, I would like to thank my students, whose interest and enthusiastic responses encouraged me to write it all down. I am particularly grateful to Crista Coles, Joe Cary, and James Gill, who worked most of the exercises in the first 10 chapters and provided invaluable feedback on the wording and presentation. I would also like to thank Ron Gruca, whose probing and incessant questions caused me to rethink some of the assumptions made in the first edition and brought about this version. Ron has provided yet more evidence that we learn at least as much from our students as they learn from us.

Mark Snavelly and his students at Carthage College used a preliminary version of the text and provided additional suggestions and encouragement. The idea for the chapter on bifurcations was entirely Mark's. Later refinements in this chapter were suggested by Roger Kraft. All errors and misrepresentations were added by the author. Roger Kraft and George Day deserve special recognition for their careful reading and correction of the entire manuscript for the first edition. Thanks are due to Ron Harrell and Jim Sandefur for reading the first edition carefully and providing a list of suggestions for improvement. Finally, I would like to express my gratitude for the support of my friends, family, and especially my wife, without which this work would not be possible.

The graphics in this book were created on a NeXT computer using *Mathematica*® and the draw program, which comes bundled with the NeXT operating system. Typesetting was accomplished using AMS-LaTeX and the AMSFonts.

Readers of this text are encouraged to contact me with their comments, suggestions, and questions. I would be very happy to hear what you think I did well and what I could do better. My e-mail address is rholmgre@alleg.edu and a full mailing address is found on the copyright page.

Richard A. Holmgren

List of Symbols

An effort has been made to use symbols and function names consistently throughout the text. The symbols used in the text are listed in the table below, along with their definitions and the page on which they are first encountered.

f^n	the nth iterate of the function f	p. 1
$\mathbb{R}$	the set of real numbers	p. 9
$f(A)$	the image of the set A	p. 10
$f^{-1}(A)$	the inverse image of the set A	p. 10
$N_\epsilon(x)$	the neighborhood of x with radius ϵ	p. 21, 110
$W^s(p)$	the stable set of the periodic point p	p. 35
$W^s(\infty)$	the stable set of infinity	p. 35
Λ_n	the set of points that remain in $[0,1]$ after n iterates of $h(x) = rx(1-x)$	p. 70
Λ	the set of points that remain in $[0,1]$ under iteration of $h(x) = rx(1-x)$	p. 70
Γ	a Cantor set	p. 73
Σ_2	the sequence space of 0's and 1's	p. 109
$d[x,y]$	the distance between the points x and y	p. 110

σ	the shift map	p. 114
S^1	a circle	p. 124
$N_f(x)$	Newton's function for the function f	p. 129
$\mathbb{C}$	the set of complex numbers	p. 167
S^2	the unit sphere in $\mathbb{R}^3$	p. 179
$\arg(z)$	the argument of the complex number z	p. 169
K_c	the filled Julia set	p. 194

Contents

Preface   **vii**
    Rationale for the new edition vii
    How to use this book viii
    Acknowledgments ix

List of Symbols   **xi**

1. Introduction   **1**
    1.1. Phase Portraits 5
    Exercise Set 1 7

2. A Quick Look at Functions   **9**
    Exercise Set 2 17

3. The Topology of the Real Numbers   **21**
    Exercise Set 3 28

4. Periodic Points and Stable Sets   **31**
    4.1. Graphical Analysis 36
    Exercise Set 4 38

5. Sarkovskii's Theorem   **41**
    Exercise Set 5 45

6. Differentiability and Its Implications   **47**
    Exercise Set 6 54

7. **Parametrized Families of Functions and Bifurcations**   59
      Exercise Set 7 67

8. **The Logistic Function Part I: Cantor Sets and Chaos**   69
      8.1. A First Look at the Logistic Function when $r > 4$ 70
      8.2. Cantor Sets 73
      8.3. Chaos and the Dynamics of the Logistic Function 76
      8.4. A Few Additional Comments on Cantor Sets 84
      Exercise Set 8 84

9. **The Logistic Function Part II: Topological Conjugacy**   87
      Exercise Set 9 92

10. **The Logistic Function Part III: A Period-Doubling Cascade**   95
      Exercise Set 10 104

11. **The Logistic Function Part IV: Symbolic Dynamics**   109
      11.1. Symbolic Dynamics and Metric Spaces 109
      11.2. Symbolic Dynamics and the Logistic Function 118
      Exercise Set 11 122

12. **Newton's Method**   127
      12.1. Newton's Method for Quadratic Functions 133
      12.2. Newton's Method for Cubic Functions 138
      12.3. Intervals and Rates of Convergence 145
      Exercise Set 12 147

13. **Numerical Solutions of Differential Equations**   153
      Exercise Set 13 163

14. **The Dynamics of Complex Functions**   167
      14.1. The Complex Numbers. 167
      14.2. Complex Functions 170
      14.3. The Dynamics of Complex Functions 174
      14.4. The Riemann Sphere 178
      14.5. Newton's Method in the Complex Plane 182
      Exercise Set 14 188

15. **The Quadratic Family and the Mandelbrot Set**   193
      15.1. Generating Julia and Mandelbrot Sets on a Computer . . 199
      Exercise Set 15 200

Appendix. *Mathematica* Algorithms   203
      A.1. Iterating Functions 204
      Finding the Value of a Point Under Iteration 204
      Tables of Iterates 204

	Controlling the Precision of the Computations	205
	Graphing Iterated Functions	206
A.2.	Graphical Analysis	206
A.3.	Bifurcation Diagrams	208
A.4.	Julia Sets	209
A.5.	The Mandelbrot Set	211
A.6.	Stable Sets of Newton's Method	212

References 215

    Dynamical Systems 215
    General Interest Books on Dynamics 217
    Topics in Mathematics 218
    Computer Programs and Algorithms 219

Index 221

1
Introduction

A discrete dynamical system can be characterized as a function that is composed with itself over and over again. For example, consider the function $f(x) = -x^3$. If we compose f with itself, then we get

$$f^2(x) = (f \circ f)(x) = -(-x^3)^3 = x^9.$$

Iterating the process we get

$$f^3(x) = (f \circ f \circ f)(x) = (f \circ f^2)(x) = -(x^9)^3 = -x^{27},$$
$$f^4(x) = (f \circ f \circ f \circ f)(x) = (f \circ f^3)(x) = -(-x^{27})^3 = x^{81},$$
$$\vdots$$
$$f^n(x) = (f \circ f^{n-1})(x) = (-1)^n x^{3^n} \quad \text{where } n \text{ is a natural number.}$$

We would like to answer the question, "Given a real number x, what is $\lim_{n \to \infty} f^n(x)$?" More generally, we ask, "What properties does the sequence $x, f(x), f^2(x), f^3(x), \ldots$ have?" By the notation $f^n(x)$, we mean f composed with itself $n-1$ times, not the nth power of f or the nth derivative of f. We call the behavior of points under iteration of the function the dynamics of the function.

Let us visualize these questions another way. Suppose that each of us lives someplace on the real line and that our address is given by the real number on which our apartment is set. For example, my address might be 2. Each year the government has decreed that I must move to a new apartment whose address can be found by cubing my present address and then finding the additive inverse. That is, I apply the function $f(x) = -x^3$ to my present address to find my new address. So, since I am currently residing at 2, next year I will be living at $f(2) = -2^3 = -8$. The year after I will be living at $f(-8) = f^2(2) = -(-8)^3 = 512$, and after n years I will be living at $f^n(2) = (-1)^n 2^{3^n}$. Even if I live to be a very old man, it is clear that I will never live in the same place twice. In fact, each year the absolute value of my address will become larger, and I will move from one side of zero to the other.

Hence, if we start at the point 2, then $f^n(2)$ grows without bound in absolute value and oscillates from one side of 0 to the other. What happens if we start someplace else? Suppose that we begin at the point $\frac{1}{2}$. Then the next year we will be at $f(\frac{1}{2}) = -\frac{1}{8}$. The following year we will be at $f(-\frac{1}{8}) = f^2(\frac{1}{2}) = \frac{1}{512}$, and after n years we will be at $f^n(\frac{1}{2}) = (-1)^n (\frac{1}{2})^{3^n}$. So each year we will move from one side of 0 to the other, but over time the move will not be very far.

In general, our address after n years will be $f^n(x_0) = (-1)^n (x_0)^{3^n}$, where x_0 is our first address. If $|x_0| > 1$, we see that

$$\lim_{n \to \infty} |f^n(x_0)| = \lim_{n \to \infty} |x_0|^{3^n} = \infty.$$

In absolute value, $f^n(x_0)$ will increase without bound and the factor $(-1)^n$ will cause $f^n(x_0)$ to oscillate from one side of 0 to the other. On the other hand, if $0 < |x_0| < 1$, then $\lim_{n \to \infty} f^n(x_0) = 0$. So while $f^n(x_0)$ will still move from one side of 0 to the other, its value will approach 0 as n gets larger. In terms of choosing apartments, an apartment whose address is less than 1 in absolute value would be preferable to an apartment whose address is larger than 1 since, as time passes, we won't have to move so far each year.

It remains to see what happens to -1, 0, and 1. Note that $f(-1) = 1$ and $f(1) = -1$. Hence, -1 and 1 form a periodic cycle. Thus, if we were living at 1 and could work out a deal with the person living at -1, then we could probably leave some of our belongings at each address since we'd be back every other year, a luxury indeed. Of course, if we lived at 0, then we'd not have to move at all, since 0 is fixed by f: a stable and perhaps boring existence.

To recap, a discrete dynamical system consists of a function and its iterates. Given a dynamical system, we would like to know where each each point goes as we iterate the function and what route it takes to get there. We have seen a number of possibilities: points that head off towards infinity, others that get close to 0, a pair of points that oscillate back and forth periodically, and a point that doesn't go anywhere at all. As we continue our investigations, the list of possibilities will grow.

Before we go on however, it is useful to look more closely at the motivation behind the study of discrete dynamical systems. In the preceding whimsical example, we used the function $f(x) = -x^3$ to determine the location of our next apartment. By iterating the function, we were able to find out where we would be in any future year. Functions and their iterates can be used to model more practical problems as well.

EXAMPLE 1.1.
Suppose that we wish to create a mathematical model describing the size of a population of rabbits living in the open fields behind my house. Suppose also that empirical evidence suggests that a small initial population will increase by approximately 10% each year. Let's assume that we start with x_0 rabbits and denote the population in the nth year by x_n. We wish to determine how many rabbits there will be in n years, or equivalently, we wish to determine the value of x_n.
Clearly,

$$x_1 = x_0 + .1x_0 = 1.1x_0$$
$$x_2 = x_1 + .1x_1 = 1.1x_1$$
$$\vdots$$
$$x_{i+1} = x_i + .1x_i = 1.1x_i.$$

Thus, $x_{i+1} = p(x_i)$ where $p(x) = 1.1x$. So,

$$x_1 = p(x_0)$$
$$x_2 = p(x_1) = (p \circ p)(x_0) = p^2(x_0)$$
$$x_3 = p(x_2) = (p \circ p^2)(x_0) = p^3(x_0)$$
$$\vdots$$
$$x_n = p^n(x_0).$$

Therefore, the population of rabbits after n years is determined by applying the function $p(x) = 1.1x$ to x_0 n times. A simple calculation shows that $p^n(x) = (1.1)^n x$. Thus, if we start with more than one rabbit, the population grows and keeps growing forever. For example, an initial population of 8 rabbits swells to a population of $(1.1)^{10} 8$, which is about 21 rabbits in 10 years. Looking further into the future, we see that according to this model, the same 8 rabbits swell to a population of approximately 54 in 20 years, 939 in 50 years, and 110,245 in 100 years. Given the small size of the field behind my house, this last estimate is clearly too large. While the dynamics of this model are easy to understand—each iterate grows by 10% over the previous iterate—the model's long-term predictive power is limited since it predicts that the population continues to grow without bound. □

In general, models that iterate a function of the form $p(x) = rx$ are called exponential models. As we demonstrated, exponential models have limited predictive power in population problems since as time passes the predicted population becomes so large that it is no longer realistic. A more sophisticated model for population, which takes into account the limits on growth, uses the logistic function, $h(x) = rx(1-x)$.

Models using the logistic function assume there is an absolute limit on the size of the population and designate the actual size of the population as a fraction of the limit. Hence, the size of any population is denoted by a number in the interval $[0, 1]$. For example, .25 indicates that the population is 25% of the limit population. If x_0 is the population in the first time period, then the population in the next time period is $h(x_0) = rx_0(1-x_0)$. The factor $(1-x)$ distinguishes models using the logistic function from exponential models. As x approaches 1, this factor approaches 0. Thus, as x becomes larger the population grows at a slower rate. If x is large enough, then the population declines. In the following example, we apply this model to the problem studied in Example 1.1.

EXAMPLE 1.2.
Let's reconsider the rabbits in the field behind my house and suppose we have determined that the limit population is 1000. That is, when there are 1000 rabbits, then they are so crowded and they so overburden the ecosystem that it is no longer able to support rabbits and the population dies out. Now consider the equation $h(x) = 1.112x(1-x)$. Recall that in this model a population of 100 is represented as $\frac{100}{1000}$ or .1 and a population of 1000 is represented as $\frac{1000}{1000}$ or 1. Clearly, $h(1) = 0$ so the model does predict that if the population reaches 1000 rabbits, then the entire colony dies out. Note also that $h(.9) \approx .1$ so that in one year the model predicts that a large initial population of 900 rabbits plummets to 100.

Now, as in Example 1.1, let's assume that we start with 8 rabbits, that is, $x_0 = .008$. Applying $h(x)$ we find that

$$x_1 = h(.008) \approx .009$$
$$x_2 = h^2(.008) \approx .01$$
$$x_3 = h^3(.008) \approx .011$$
$$x_4 = h^4(.008) \approx .012.$$

As the number of iterates increases, we find that

$$x_{10} = h^{10}(.008) \approx .02$$
$$x_{20} = h^{20}(.008) \approx .043$$
$$x_{100} = h^{100}(.008) \approx .101.$$

Thus, the population still grows by approximately 10% each year when the size of the population is small, but the rate of growth slows as the population gets larger. In the exercises, the reader is asked to experiment with other initial populations over time and to describe the dynamics of the population size as predicted by this model. □

We will see in the following chapters that in general the dynamics of the logistic function are much harder to understand than the dynamics of exponential models. In fact, much of this book is devoted to understanding the dynamics of this function.

1.1. Phase Portraits

Phase portraits are frequently used to graphically represent the dynamics of a system. A phase portrait consists of a diagram representing possible beginning positions in the system and arrows that indicate the change in these positions under iteration of the function. It is best understood by looking at a few examples.

EXAMPLE 1.3.
Let $f(x) = x^2$. The dynamical system we are considering consists of the domain of f and the function itself. The domain is the set of real numbers, which we represent by a line. Note that 0 and 1 are fixed, that is, $f(0) = 0$ and $f(1) = 1$. We indicate this in our phase portrait by dots at 0 and 1. We notice that if $0 < x < 1$, then $f^n(x)$ tends towards 0 as n becomes larger, and if $x > 1$, then $f^n(x)$ tends towards infinity as n becomes larger. To represent this, we draw an arrow from 1 towards 0 and another arrow from 1 towards positive infinity. Now what happens if x is less than 0? The point -1 becomes fixed at 1 since $f(-1) = 1$. This implies $f^n(-1) = 1$ for all n larger than or equal to 1. This is shown by the arrow from -1 to 1. Points that lie between -1 and 0 are mapped into the interval $(0, 1)$ and then they move towards 0 under iteration of the function. Similarly, points which are to the left of -1 are mapped to the right of 1 and then move towards infinity under iteration of the map. We represent this by arrows from the negative portion of the line to the positive portion. All of this information is encoded in the phase portrait shown in Figure 1.1. □

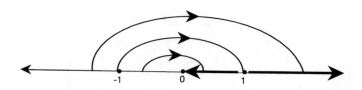

FIGURE 1.1. The phase portrait of $f(x) = x^2$.

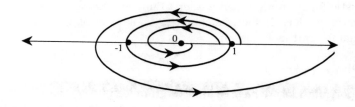

FIGURE 1.2. The phase portrait of $f(x) = -x^3$.

EXAMPLE 1.4.
Consider the function $f(x) = -x^3$. We recall from our earlier discussion that 0 is fixed and the point 1 goes to -1 and then returns to 1 on the second iteration. Points that are greater than 1 in absolute value oscillate from one side of zero to the other under iteration of f and grow ever larger in absolute value. Points that are less than 1 in absolute value oscillate from one side of zero to the other under iteration of f and grow ever smaller in absolute value. All of this is encoded in the phase portrait shown in Figure 1.2. □

The reader is given an opportunity to explore and develop more phase portraits in the exercises.

Exercise Set 1

1.1 Suppose $h(x) = rx(1-x)$.

a) Let $r = 2$ and calculate $h^n(.25)$ for $n = 0, 1, 2, 3$, and 4 and $n = 101, 102, 103, 104$. Can you determine $\lim_{n \to \infty} h^n(.25)$?

b) Let $r = 2$ and calculate $h^n(.2345)$ for $n = 0, 1, 2, 3$, and 4 and $n = 101, 102, 103, 104$. Can you determine $\lim_{n \to \infty} h^n(.2345)$?

c) Let $r = 3.1$ and calculate $h^n(.25)$ for $n = 0, 1, 2, 3$, and 4 and $n = 101, 102, 103, 104$. Can you determine $\lim_{n \to \infty} h^n(.25)$?

d) Let $r = 3.1$ and calculate $h^n(.2345)$ for $n = 0, 1, 2, 3$, and 4 and $n = 101, 102, 103, 104$. Can you determine $\lim_{n \to \infty} h^n(.2345)$?

e) Let $r = 4$ and calculate $h^n(.25)$ for $n = 0, 1, 2, 3$, and 4 and $n = 101, 102, 103, 104$. Can you determine $\lim_{n \to \infty} h^n(.25)$?

f) Let $r = 4$ and calculate $h^n(.2345)$ for $n = 0, 1, 2, 3$, and 4 and $n = 101, 102, 103, 104$. Can you determine $\lim_{n \to \infty} h^n(.2345)$?

1.2 Describe Newton's method for approximating the zeros of a function as a dynamical system. (You should be able to find Newton's method in your calculus textbook.) What is the significance of $\lim_{n \to \infty} f^n(x)$ in this case?

1.3 Draw the phase portraits for the following functions:

a) $f(x) = x^{1/3}$

b) $g(x) = 2 \arctan x$

c) $r(x) = \frac{1}{x}$

d) $C(x) = \cos(x)$

1.4 Explore the logistic model developed in Example 1.2 by experimenting with a variety of initial populations. Explain why it only makes sense to choose initial populations between 0 and 1000. Describe the expected growth of the population for a variety of initial populations. Represent the behavior of the population using a phase portrait.

1.5 On page 4, we were introduced to the logistic function as a model for population growth. Recall that the logistic function is defined as $h(x) = rx(1-x)$. We stated that if x is large enough, then the population declines. Suppose $r = 2.5$. Find a point x_0 in $[0, 1]$ so that $h(x) \leq x$ if and only if $x_0 \leq x$. What happens to the size of the population over time if the initial population density is x_0? What if the initial density is a little larger than x_0? a little smaller? Answer the same questions for $r = 3.4$ and $r = 1$.

1.6 AN INVESTIGATION: Let $h(x) = rx(1 - x)$. Investigate the changing behavior of the finite sequence $h^{500}(.1), h^{501}(.1), \ldots, h^{535}(.1)$ as r varies from 3 to 3.6. You might begin by looking at the r values 3, 3.1, 3.2, 3.3, 3.4, 3.5, and 3.6. Between which values do you see a qualitative change in behavior? If you see a change between $r = 3.4$ and $r = 3.5$, then look at the behavior at the midpoint $r = 3.45$. Continue to focus on smaller and smaller intervals around changes in behavior. You should find that a change in the r value as small as .005 can be significant. Describe these changes and the places they occur to the best of your ability.

1.7 AN INVESTIGATION: If you have been using a computer to investigate the behavior of a function under iteration, then you should be aware that the program you are using rounds off to a fixed number of significant digits after each iteration. You might also have wondered whether or not this rounding affected the outcome of your investigations. Write a short program in which the number of significant digits used in calculations can be chosen by the user. Such a program for *Mathematica* can be found in the appendix. Investigate the behavior of points in $[0, 1]$ under iteration of $h(x) = rx(1 - x)$ when the computations are done with a precision of 8 significant digits and again when they are done with 99 significant digits using various values of r between 3 and 4. (A good place to start is to consider what happens when $r = 3$, $r = 3.5$, and $r = 4$.) Does the number of significant digits matter? For which values of r?

2
A Quick Look at Functions

Before we begin our discussion of dynamics, it is necessary to review a few definitions. We review the terms function, one-to-one, onto, continuous, and inverse function in this chapter. A good, working knowledge of these terms is fundamental to understanding the material in subsequent chapters. We also introduce homeomorphisms, which will be important in Chapter 9, and complete the chapter by reviewing the Intermediate Value Theorem.

DEFINITION 2.1. *A function is a rule that assigns each element of one set to a unique element of a second set. The first set is called the domain of the function, and the second set is called the codomain. The set of elements in the codomain that have an element of the domain assigned to them is called the range of the function. We use the notation $f : D \to C$ to indicate a function f with domain D and codomain C. The notation $f : D \to D$ indicates that the domain and codomain of the function are the same set.*

We often refer to functions as *maps* or *mappings*. Note that a function isn't necessarily a map from one set of real numbers to another. While functions of the real numbers are certainly important, they are by no means the only examples. There are many important functions that are not functions of the real numbers; we will be introduced to one such function in Chapter 11 when we study symbolic dynamics. When we wish to refer to the set of real numbers we use the symbol $\mathbb{R}$. Thus, the notation $g : \mathbb{R} \to \mathbb{R}$

indicates a function g whose domain and codomain are both the set of real numbers.

If $f: D \to C$ and $A \subset D$, then $f(A)$ is defined to be the subset of C containing all the elements of the form $f(a)$ where a is in A. That is,

$$f(A) = \{c \text{ in } C \mid \text{there is } a \text{ in } A \text{ satisfying } f(a) = c\}.$$

The set $f(A)$ is often called the *image* of A under f. Notice that if $f: D \to C$, then $f(D)$ is the range of f. If $B \subset C$, then $f^{-1}(B)$ is called the *inverse image* or *preimage* of B and consists of all elements of D whose image is contained in B. That is,

$$f^{-1}(B) = \{x \text{ in } D \mid f(x) \text{ is in } B\}.$$

EXAMPLE 2.2.
Let $f: \mathbb{R} \to \mathbb{R}$ be defined by $f(x) = x^2$. Then

$$f([-1, 2]) = [0, 4] \text{ and } f^{-1}((1, 4]) = [-2, -1) \cup (1, 2].$$

Note that

$$f(f^{-1}([1, 4])) = [1, 4], \text{ but } f^{-1}(f([1, 4])) = [-4, -1] \cup [1, 4].$$

Note also that the use of the notation f^{-1} does not necessarily imply that f is an invertible function. □

DEFINITION 2.3. *A function is one-to-one if there is exactly one element of its domain assigned to each element of its range.*

PROPOSITION 2.4. *The function f is one-to one if and only if the statement $f(x) = f(y)$ implies that $x = y$.*

The "if and only if" in this proposition means that the two statements, "$f: D \to C$ is one-to-one" and "$f(x) = f(y)$ implies $x = y$" are equivalent. In other words, if one of the statements is true, then the other is true; if one of the statements is false, then the other is false. The proof of Proposition 2.4 is not hard and is left as an exercise.

Proposition 2.4 provides an easy way to test whether or not a function is one-to-one. For example, consider the function $f: \mathbb{R} \to \mathbb{R}$ defined by $f(x) = x^2$. Clearly, f is not one-to-one since $f(1) = 1 = f(-1)$. On the other hand, the function $g: [0, \infty) \to \mathbb{R}$ defined by $g(x) = x^2$ is one-to-one since if $g(x) = g(y)$, then $x^2 = y^2$, $x \geq 0$, and $y \geq 0$ imply $x = y$. Note that even though the functions f and g have the same rule, they have different domains and, hence, different properties.

A graphical test for determining whether functions of the real numbers are one-to-one is the *horizontal line test*. A function is one-to-one if and only if every horizontal line crosses the graph of the function at most once.

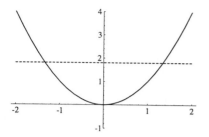
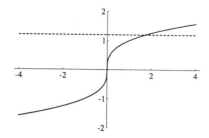

FIGURE 2.1. The graphs of $f(x) = x^2$ and $p(x) = x^{1/3}$.

By looking at the graphs of $f(x) = x^2$ and $p(x) = x^{1/3}$, we see that f is not one-to-one and p is. The graphs of f and p are shown in Figure 2.1. In the exercises, we ask the reader to explain why the horizontal line test works.

DEFINITION 2.5. *A function is onto if each element of the codomain has at least one element of the domain assigned to it. In other words, a function is onto if the range equals the codomain.*

When determining whether or not a function is onto we must know what the codomain is. For example, the function $f : \mathbb{R} \to \mathbb{R}$ defined by $f(x) = e^x$ is not onto. There is no point in the domain that is mapped to any number in the interval $(-\infty, 0]$. On the other hand, the function $g : \mathbb{R} \to (0, \infty)$ defined by $g(x) = e^x$ is onto. Of course, every function can be made into an onto function by restricting its codomain to its range.

What can we say about the graph of an onto function whose domain and codomain are subsets of the real numbers? If we consider the codomain as a subset of the y-axis, then it is clear that every horizontal line that passes through an element of the codomain must also pass through the graph of an onto function. This geometric description is useful when we are trying to construct examples of functions with specific properties.

EXAMPLE 2.6.
Construct an example of a function $f : [1, 3] \to [2, 4]$ that is onto but not one-to-one.

Let us first determine what properties the graph of f must have. Since all the values of f must lie in the codomain, the graph lies between the lines $y = 2$ and $y = 4$. As the domain of f is $[1, 3]$, a vertical line intersects the graph if and only if it passes through a point on the x-axis between and including 1 and 3. So the graph of the function lies in the box with corners $(1, 2)$, $(1, 4)$, $(3, 4)$, and $(3, 2)$, as shown in Figure 2.2.

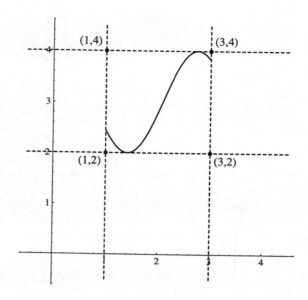

FIGURE 2.2. The graph of a function, $f : [1, 3] \to [2, 4]$, which is onto, but not one-to-one.

Since f is onto, we know that every horizontal line that passes through the codomain (thought of as a subset of the y-axis) must intersect the graph of the function. Also, since f is not one-to-one there must be at least one horizontal line through the codomain that strikes the graph two or more times. Obviously, there are many graphs that satisfy these criteria. One example is shown in Figure 2.2. □

While it is often sufficient to think of continuous functions as functions whose graphs are unbroken lines, it is essential to have a more precise definition. In particular, if a function's domain or codomain is not an interval of the real line, then the graph of the function is not always so easy to draw and not always informative when it can be drawn. For example, if $\mathbb{Z}$ is the set of integers, then the function $f : \mathbb{Z} \to \mathbb{Z}$ defined by $f(z) = 2z$ is continuous, but its graph is a set of discrete points.

DEFINITION 2.7. *Let $f : D \to C$ and x_0 be a point in D. Then f is continuous at x_0 if for every positive number ϵ, there exists a positive number δ such that*

$$\text{if } x \text{ is in } D \text{ and } |x_0 - x| < \delta, \text{ then } |f(x_0) - f(x)| < \epsilon.$$

A function is continuous if it is continuous at each point of its domain.

FIGURE 2.3. An illustration of continuity at the point x_0.

To understand this definition, it is helpful to think of $|x - y|$ as the distance from x to y. Then $|x - y| < \epsilon$ simply means that the distance from x to y on the real line is smaller than ϵ. When contemplating the definition of continuity, visualize x_0 and $f(x_0)$ as points in the domain and codomain of f, as shown in Figure 2.3. Now draw a circle around $f(x_0)$ with a radius of ϵ and its center at $f(x_0)$. If f is continuous at x_0, then we can draw a circle around x_0 whose radius δ is so small that whenever x is in the circle around x_0 and in the domain of f, then $f(x)$ is in the circle around $f(x_0)$. Note that x is in the circle of radius δ around x_0 if and only if $|x - x_0| < \delta$. That is, x is in the circle of radius δ around x_0 if and only if the distance from x to x_0 is less than δ. Similarly, $f(x)$ is in the circle around $f(x_0)$ of radius ϵ if and only if $|f(x) - f(x_0)| < \epsilon$.

EXAMPLE 2.8.
Let $\mathbb{Z}$ be the set of integers. We use the definition to show that the function $f : \mathbb{Z} \to \mathbb{Z}$ defined by $f(z) = 2z$ is continuous.

Suppose a is any integer and ϵ is any number greater than 0. We must find a number δ that is greater than 0 and such that $|f(z) - f(a)| < \epsilon$ whenever z is an integer satisfying $|z - a| < \delta$. Let's try $\delta = \frac{1}{2}$. If z is an integer and $|z - a| < \frac{1}{2}$, then z must equal a. Hence, $|f(z) - f(a)|$ is equal to 0, which is less than ϵ. □

Finding a δ for the given ϵ in Example 2.8 was straightforward. However, when demonstrating continuity at a point, it is not always as easy to find an appropriate δ. We consider a more complicated case in Example 2.9. The Triangle Inequality is used to analyze this example.

The Triangle Inequality needn't be mysterious. As most of us realize, it is further to go from New York to Chicago by way of Los Angeles than by traveling directly; this is not surprising. Essentially, the Triangle Inequality tells us that the distance from point a to point b is less than or equal to the distance from a to point c plus the distance from c to b. Recalling that the value $|a - b|$ represents the distance from point a to point b, we can express the Triangle Inequality mathematically by

$$|a - b| \leq |a - c| + |c - b| \tag{2.1}$$

where a, b, and c represent real numbers. If we let $a = x + c$ and $b = c - y$ in formula (2.1), then we get

$$|x + y| \leq |x| + |y|.$$

This is the form of the Triangle Inequality that is most useful to us; we use it in virtually every chapter of this text. The reader is well advised to take the time to prove it and learn how to use it. Exercise 2.8 at the end of this chapter provides hints for proving the inequality and practice using it. Our first use of the Triangle Inequality is in the following example.

EXAMPLE 2.9.
Show that the function $g(x) = x^2$ is continuous on the real numbers.

Let a be any real number. We must show that g is continuous at a. We consider the case where a is not 0. Let ϵ be any positive real number. We must find a positive number δ such that whenever x is a real number satisfying $|x - a| < \delta$, then $|g(x) - g(a)| = |x^2 - a^2| < \epsilon$. To accomplish this, we choose δ so that $\delta < |a|$, $\delta < \frac{\epsilon}{|3a|}$, and $\delta > 0$. This is possible since both ϵ and $|a|$ are positive numbers. If $|x - a| < \delta$, then using the Triangle Inequality, we find that

$$|x| = |x - a + a| \leq |x - a| + |a| < |a| + \delta.$$

It follows that if $|x - a| < \delta$, then

$$\begin{aligned}|g(x) - g(a)| = |x^2 - a^2| &= |x - a||x + a| \\ &< \delta|x + a| \\ &\leq \delta(|x| + |a|) \\ &< \delta(|a| + \delta + |a|) \\ &< \delta|3a| \\ &< \frac{\epsilon}{|3a|}|3a| = \epsilon.\end{aligned}$$

Hence, $|g(x) - g(a)| < \epsilon$ when $|x - a| < \delta$ as desired. Thus, g is continuous at each nonzero real number. Note that the Triangle Inequality was used twice in the preceding sequence of inequalities: once in going from the second line to the third and a second time in going from the third line to the fourth. We leave the proof that g is continuous at 0 as an exercise. □

Note that in proving that g was continuous at a in the previous example, our choice of δ depended on the value of *both* a *and* ϵ. The dependence of δ on both the point in question and ϵ is the rule rather than the exception.

The lemma that follows provides another way of describing continuity. Essentially, this lemma states that $f(x)$ is continuous at the number p if and only if for every interval containing $f(p)$ there is an interval containing p whose image is contained in the first interval.

LEMMA 2.10. *The function $f(x)$ is continuous at p if and only if for each c and d satisfying $c < f(p) < d$, there is $\delta > 0$ such that if x is in the domain of f and x is in $(p - \delta, p + \delta)$, then $c < f(x) < d$.*

PROOF. We begin by assuming that $f(x)$ is continuous at p and that $c < f(p) < d$. We must find $\delta > 0$ so that if x is in $(p - \delta, p + \delta)$ and $f(x)$ is defined, then $c < f(x) < d$.

Choose ϵ to be the smaller of $d - f(p)$ and $f(p) - c$. Since $f(x)$ is continuous at p, there is $\delta > 0$ such that if $|x-p| < \delta$, then $|f(x)-f(p)| < \epsilon$. But $|x - p| < \delta$ is equivalent to $-\delta < x - p < \delta$ or $p - \delta < x < p + \delta$. The latter statement is true if and only if x is in $(p - \delta, p + \delta)$. Similarly, $|f(x) - f(p)| < \epsilon$ is equivalent to

$$f(p) - \epsilon < f(x) < f(p) + \epsilon. \tag{2.2}$$

Since $f(p) - c \geq \epsilon$ and $\epsilon \leq d - f(p)$, inequality (2.2) implies

$$f(p) - (f(p) - c) < f(x) < f(p) + (d - f(p))$$

or

$$c < f(x) < d.$$

It follows that if x is in $(p - \delta, p + \delta)$, then $c < f(x) < d$.

Now assume that for any c and d satisfying $c < f(p) < d$, we can find $\delta > 0$ such that if x is in $(p - \delta, p + \delta)$, then $c < f(x) < d$. We must show that $f(x)$ is continuous at p. That is, for any $\epsilon > 0$ there is $\delta > 0$ so that if $|x - p| < \delta$, then $|f(x) - f(p)| < \epsilon$. By arguments similar to the ones used in the preceding paragraph, this is equivalent to showing that there is $\delta > 0$ so that if x is in $(p - \delta, p + \delta)$, then $f(p) - \epsilon < f(x) < f(p) - \epsilon$, and this follows immediately from our assumption. The reader is encouraged to fill out the details of this half of the proof. □

The astute reader may have noticed that our definition of continuity only makes sense if the notion of a distance is defined on the sets D and C. Given our current knowledge, this means that D and C must be subsets of the real numbers. When we look at metric spaces in Chapter 11, we will extend the definition of continuity to functions whose domain and codomain may be more general sets.

The Intermediate Value Theorem, which we introduce next, is a useful property of continuous functions. An application of the Intermediate Value Theorem is demonstrated in Exercise 2.18. Further applications are found in subsequent chapters.

THEOREM 2.11. THE INTERMEDIATE VALUE THEOREM. *Let f be a continuous function defined on the interval $[a,b]$. If p is a number between $f(a)$ and $f(b)$, then there exists c in $[a,b]$ such that $f(c) = p$. That is, if $f(a) \leq p \leq f(b)$ or $f(b) \leq p \leq f(a)$, then there is c in $[a,b]$ such that $f(c) = p$.*

A proof of the Intermediate Value Theorem can be found in introductory analysis textbooks such as those listed in the references. We note that the condition that f be continuous is *necessary*. We demonstrate this in the following example.

EXAMPLE 2.12.
Let $f : [0,2] \to \mathbb{R}$ be defined by $f(x) = \begin{cases} 1, & \text{for } x < 1 \\ 3, & \text{for } x \geq 1 \end{cases}$. Then $f(0) = 1$ and $f(2) = 3$. Even though 2 is between $f(0)$ and $f(2)$ there is no number c in $[0,3]$ such that $f(c) = 2$. This does not contradict the Intermediate Value Theorem since $f(x)$ is not continuous at 1. (The reader may wish to verify that f is indeed not continuous at 1.) □

DEFINITION 2.13. *Let $f : D \to C$ be a function with range $f(D)$. The function $g : f(D) \to D$ is an inverse of f if $(g \circ f)(x) = x$ for all x in D. If f has an inverse, then it is usually denoted f^{-1}, and we say that f is invertible.*

The following proposition is well-known; its proof is left as an exercise.

PROPOSITION 2.14. *A function is invertible if and only if it is one-to-one.*

DEFINITION 2.15. *If a function is one-to-one, onto, and continuous, and its inverse is continuous, then the function is a homeomorphism. In this case, we say the domain and codomain are homeomorphic to one another.*

Homeomorphisms play an important role in mathematics in general and in dynamical systems in particular. When two mathematical spaces or structures are homeomorphic they are, in some sense, the same. Mathematically, we describe this by saying that they have the same topological properties. The intervals $[1,2]$ and $[3,5]$ are homeomorphic since $g : [1,2] \to [3,5]$ defined by $g(x) = 2x + 1$ is a homeomorphism. (The reader should check that g is in fact one-to-one, onto, continuous, and has a continuous inverse.) However, the intervals $[1,2]$ and $(3,5)$ are not homeomorphic. That is, there is no function $f : [1,2] \to (3,5)$ that is one-to-one, onto, continuous, and has a continuous inverse. This is demonstrated in Exercise 2.18. In Chapter 3, we will see that the topologies of $[1,2]$ and $(3,5)$

are different as well. Specifically, [1, 2] contains all of its limit points while (3, 5) does not.

Note that in all of the examples in this book, if a function is one-to-one, onto, and continuous, then it is a homeomorphism. *This is not true in general.* There are functions that are one-to-one, onto, and continuous but that are not homeomorphisms. The interested reader may consult *General Topology* by S. Willard or a similar reference for further details.

Exercise Set 2

2.1 Let $f(x) = x^2$ and $S(x) = \sin x$.

    a) Find $f([0, 1])$.

    b) Find $f^{-1}(f([0, 1]))$ and $f(f^{-1}([0, 1]))$. Why aren't these two sets the same?

    c) Find $S([0, \frac{\pi}{2}])$ and $S^{-1}(S([0, \frac{\pi}{2}]))$.

2.2 Let $g : \mathbb{R} \to \mathbb{Z}$ be the step function defined such that $g(x)$ is the largest integer less than or equal to x. ($\mathbb{Z}$ denotes the set of integers.) Calculate $g([\frac{1}{2}, 10.1])$ and $g^{-1}(\{2, 4\})$.

2.3 a) Let A be a subset of the domain of the function f. Prove $f^{-1}(f(A)) \supset A$ and show by example that it may happen that $f^{-1}(f(A)) \neq A$.

    b) Let B be a subset of the codomain of the function f. Determine which of the following must be true and show that your answer is correct:

$$f(f^{-1}(B)) \supset B, \qquad f(f^{-1}(B)) \subset B, \qquad f(f^{-1}(B)) = B.$$

2.4 Suppose $f : [-1, 1] \to [-1, 1]$ is defined so that $f(x) = \frac{1}{2}x$.

    a) Find the range of f.

    b) Find the range of $f \circ f$.

    c) Find the range of f^n for all natural numbers n.

2.5 a) Find a function $f : [0, 2] \to [2, 4]$ so that f is one-to-one and continuous but not onto.

b) Find a function $f : [0, 2] \to [2, 4]$ so that f is onto and continuous but not one-to-one.

c) Find a function $f : [0, 2] \to [2, 4]$ so that f is one-to-one and onto but not continuous.

d) Find a function $f : [0, 2] \to [2, 4]$ so that f is a homeomorphism. That is, f is one-to-one, onto, continuous, and has a continuous inverse.

2.6 Prove Proposition 2.4.

2.7 Prove that a function of the real numbers is one-to-one if and only if every horizontal line intersects the graph of the function at most once.

•2.8 THE TRIANGLE INEQUALITY.

a) Let a and b be real numbers. Prove that $|a + b| \leq |a| + |b|$. You may use the fact that $-|a| \leq a \leq |a|$ and $-|b| \leq b \leq |b|$ or you may wish to show that $|a + b|^2 \leq (|a| + |b|)^2$.

b) Use part (a) to prove that $|x - y| \leq |x - z| + |z - y|$.

c) Prove $|x - y| \leq |x| + |y|$.

d) Prove $|x| \geq |x - y| - |y|$.

e) Prove $|x| - |y| \leq |x - y|$.

2.9 a) Justify each of the steps in the second displayed inequality of Example 2.9.

b) Prove that $g(x) = x^2$ is continuous at 0.

2.10 Assume that for any c and d satisfying $c < f(p) < d$ we can find $\delta > 0$ such that if x is in $(p - \delta, p + \delta)$ and $f(x)$ is defined, then $c < f(x) < d$. Show that $f(x)$ is continuous at p. Note that this is the second half of the proof of Lemma 2.10. A sketch of the proof is provided immediately following the statement of the lemma.

2.11 Prove Proposition 2.14.

2.12 Show that $f(x) = \frac{1}{x}$ is continuous.

2.13 a) Show that $f(x) = \begin{cases} -1 & \text{for } x < 0 \\ 1 & \text{for } x > 0 \end{cases}$ is continuous.

b) Show that $f(x) = \begin{cases} -1 & \text{for } x \leq 0 \\ 1 & \text{for } x > 0 \end{cases}$ is not continuous.

2.14 Show that every function whose domain is the integers and whose codomain is the real numbers is continuous.

•2.15 a) Show that all nonconstant linear functions are continuous.
Note: A nonconstant linear function has the form $f(x) = mx + b$ where $m \neq 0$.

b) Show that all nonconstant linear functions have continuous inverses.

c) Show that all nonconstant linear functions are homeomorphisms.

•2.16 Let $f : [a,b] \to \mathbb{R}$ be a continuous function. Suppose $f(a) = c$, $f(b) = d$, and $c < d$. Prove that $f([a,b]) \supset [c,d]$.
Hint: Use the Intermediate Value Theorem to show that every number between c and d has something mapped to it by f.

2.17 Prove that if a and b are real numbers satisfying $a < b$, then the interval $[a,b]$ is homeomorphic to $[0,1]$.
Hint: Find a linear function $f : [a,b] \to [0,1]$.

2.18 a) Let $f : [1,2] \to (3,5)$ be one-to-one and continuous. Show that f is not onto.
Hint: First show that f is strictly increasing or strictly decreasing by using the Intermediate Value Theorem.
Note: This exercise implies that [1,2] and (3,5) are not homeomorphic.

b) Let $g : (0,1) \to [1,3]$ be onto and continuous. Show that g is not one-to-one.
Hint: First show there is x_0 in $(0,1)$ such that $g(x_0) = 3$. What is the nature of g near x_0? You may wish to use the Intermediate Value Theorem as well.
Note: This exercise implies that (0,1) and [1,3] are not homeomorphic.

c) Let a, b, c, and d be real numbers satisfying $a < b$ and $c < d$. Prove that the intervals (a, b) and $[c, d]$ are not homeomorphic.

•2.19 Let $f : D \to C$ and $g : C \to E$. Prove that the following statements are true:

a) If f and g are onto, then $g \circ f$ is onto.

b) If f and g are one-to-one, then $g \circ f$ is one-to-one.

c) If f and g are continuous, then $g \circ f$ is continuous.

d) If f and g are homeomorphisms, then $g \circ f$ is a homeomorphism.

•2.20 Let $f : D \to C$ and $g : D \to C$ be continuous functions. Prove that $f + g$ and $f - g$ are continuous functions.

*2.21 Prove or disprove: If $f : \mathbb{R} \to \mathbb{R}$ is one-to-one, onto, and continuous, then f is a homeomorphism. (Prove or disprove means that you should either prove the statement or find an example of a function, $f : \mathbb{R} \to \mathbb{R}$, that is one-to-one, onto, and continuous but not a homeomorphism.)

3
The Topology of the Real Numbers

The topology of a mathematical space is its structure or the characteristics it exhibits. In calculus, we were introduced to a few topological ideas, and we will need a few more in our study of dynamics.

One of the fundamental questions of dynamics concerns the properties of the sequence x, $f(x)$, $f^2(x)$, $f^3(x), \ldots$. To discuss these properties intelligently we need to understand convergence, accumulation points, open sets, closed sets, and dense subsets. In this section, we will limit our discussion to subsets of the real numbers; we will revisit the definitions when we introduce metric spaces in Chapter 11.

DEFINITION 3.1. *Suppose U is a subset of the real numbers. Then U is open if for each x in U there is a positive number ϵ so that $|x - y| < \epsilon$ implies y is in U.*

As with the definition of continuity, it is useful to think of the ϵ in this definition as a distance. Picture U as a subset of the real line as shown in Figure 3.1, and let x be a point in U. If U is open, then we can find an $\epsilon > 0$ such that when we draw a circle around x with radius ϵ, all the real numbers contained in the circle are contained in U. This disk with radius ϵ and center x is usually called an ϵ-neighborhood or simply a neighborhood of x.

DEFINITION 3.2. *Let $\epsilon > 0$. The set $N_\epsilon(x) = \{y \text{ in } \mathbb{R} \mid |x - y| < \epsilon\}$ is an ϵ-neighborhood of x. To simplify our terminology, we will drop the ϵ*

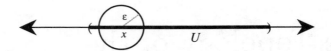

FIGURE 3.1. An open set.

and call $N_\epsilon(x)$ a neighborhood of x when it will not cause confusion.

We can think of a neighborhood of x as all the points that "live nearby" x. Note that $|x - y| < \epsilon$ if and only if y is in the open interval $(x - \epsilon, x + \epsilon)$. Consequently, $N_\epsilon(x) = (x - \epsilon, x + \epsilon)$, and we can visualize neighborhoods of x as open intervals centered at x.

The following two propositions follow from a careful examination of the definitions of open sets and neighborhoods and should be proven by the reader.

PROPOSITION 3.3. *The set U is open if and only if for each x in U there is a neighborhood of x that is completely contained in U.*

PROPOSITION 3.4. *Every neighborhood of a real number is an open set.*

In the following example, we use the definition of open to demonstrate that "open intervals" are indeed open and to show that not every set need be open.

EXAMPLE 3.5.
a) The interval (a, b) is open. To see this, let x be a point in (a, b). To show that (a, b) is open, we must find $\epsilon > 0$ so that $|y - x| < \epsilon$ implies y is in (a, b). Suppose that ϵ is the smaller of $|x - a|$ and $|x - b|$. In other words, suppose that ϵ is the distance from x to the nearest endpoint of (a, b). Since $a < x < b$, ϵ is positive. If $|x - y| < \epsilon$, then

$$|x - y| < \epsilon \leq |x - a|.$$

Since $x > a$, this implies

$$-x + a < x - y < x - a.$$

Solving this for y, we find that

$$a < y < 2x - a.$$

Hence, $a < y$. To complete the proof that y is in (a, b), we need to show that $y < b$. The proof is similar to the preceding one. The details are left as an exercise.

b) The set of rational numbers is not an open set. This follows from the fact that if x is a rational number, then no matter how small we choose the positive number ϵ to be, there is some irrational number whose distance from x is less than ϵ. We prove this formally in the exercises. □

It is important to notice the difference in the way we use ϵ in the definitions of continuity and open sets. When demonstrating that a set is open, we find a positive ϵ that works for each x. On the other hand, in the case of continuity at a point, we must find an appropriate δ for *every* positive ϵ. However, the two concepts are related, as we demonstrate with the following theorem.

THEOREM 3.6. *The function $f : \mathbb{R} \to \mathbb{R}$ is continuous if and only if the preimage of every open set is open. In other words, f is continuous if and only if for each open set U in $\mathbb{R}$ the set $f^{-1}(U)$ is open.*

Again, the proof of this theorem is not difficult, and interested readers are encouraged to prove it. Some individuals may find Lemma 2.10 useful when completing the proof.

Since there are open sets in mathematics, it's not surprising that there are closed sets as well. To define closed sets, we use the concepts of convergent sequences and accumulation points.

DEFINITION 3.7. *Let $x_1, x_2, x_3, \ldots$ be a sequence of real numbers. The sequence converges to x if for each $\epsilon > 0$, there exists N such that if $k \geq N$, then $|x - x_k| < \epsilon$. The sequence grows without bound or converges to infinity if for each M there exists N such that if $k \geq N$, then $x_k > M$.*

Once again, we can treat ϵ as a distance. If the sequence $x_1, x_2, x_3, \ldots$ converges to x, then when k is large, x_k is close to x, that is, $|x_k - x| < \epsilon$. Notice that to demonstrate that a sequence converges we must be able to find an N that works for each ϵ. The N may change as ϵ changes; it is not sufficient to find an N that works for one particular ϵ.

EXAMPLE 3.8.
a) The sequence $x_1 = .9$, $x_2 = .99$, $x_3 = .999, \ldots$ converges to 1. To see this, let $\epsilon > 0$. Choose an integer N such that $\frac{1}{10^N} < \epsilon$. Then $|1 - x_k| = \frac{1}{10^k} \leq \frac{1}{10^N} < \epsilon$ for all $k \geq N$ as desired.

b) The sequence $x_1 = 1$, $x_2 = 2$, $x_3 = 1$, $x_4 = 2$, $x_5 = 1, \ldots$ does not converge to any real number. To demonstrate this, it is sufficient to find a single ϵ so that for every N and every real number x we can find $k \geq N$ such that $|x_k - x| > \epsilon$. Set $\epsilon = \frac{1}{4}$ and let x be any real number. We claim that if x_N is such that $|x_N - x| < \frac{1}{4}$, then $|x_{N+1} - x| > \frac{1}{4}$. To see this,

24   3. The Topology of the Real Numbers

suppose that $x_N = 1$; it has to be either 1 or 2. Then $x_{N+1} = 2$ and by the Triangle Inequality,

$$\begin{aligned}|x_{N+1} - x| &= |2 - x| \\ &= |1 - (x - 1)| \\ &\geq |1| - |x - 1| > \frac{3}{4}.\end{aligned}$$

A similar result holds when $x_N = 2$. Hence, when $\epsilon = \frac{1}{4}$, there does not exist an x and N such that $|x_k - x| < \epsilon$ for all $k \geq N$. □

Another frequently encountered topological concept is that of an accumulation point.

DEFINITION 3.9. *Let S be a subset of the real numbers. Then the real number x is an accumulation point (or a limit point) of S if every neighborhood of x contains an element of S that is not x.*

Loosely speaking, a number is an accumulation point of a set if we can find points of the set that are as close to the number as we want. Note that a number need not be in a set to be an accumulation point of it. This will be demonstrated in Example 3.11b. The following proposition gives us two alternative definitions of accumulation point and is an easy consequence of the preceding definitions.

PROPOSITION 3.10. *Let S be a set. Then the following statements are equivalent.*

a) *The point x is an accumulation point of the set S. That is, every neighborhood of x contains at least one element of S that is not x.*

b) *For each $\epsilon > 0$ there exists y in S such that $0 < |x - y| < \epsilon$.*

c) *There is a sequence of points, all different from x and contained in S, that converges to x.*

The reader should prove the equivalence of the three statements in Proposition 3.10. Once this is done, we can use any one of the statements to prove assertions about accumulation points.

EXAMPLE 3.11.

a) The point 0 is an accumulation point of $(0, 1]$. This follows from the fact that if ϵ is any positive number, then either $1 < \epsilon$ or $\frac{\epsilon}{2}$ is an element of $(0, 1]$. In the first case, 1 is in $(0, 1]$ and $0 < |0 - 1| < \epsilon$. In the second case, $\frac{\epsilon}{2}$ is in $(0, 1]$ and $0 < |0 - \frac{\epsilon}{2}| < \epsilon$. Thus 0 is an accumulation point of $(0, 1)$ by part (b) of Proposition 3.10.

b) The irrational number $x = .123456789101112131415\ldots$ is an accumulation point of the set of rational numbers. To prove this, we define a sequence of rational numbers by

$$x_1 = .1$$
$$x_2 = .12$$
$$x_3 = .123$$
$$x_4 = .1234$$
$$\vdots$$

It is easy to prove that this sequence converges to x. (The reader is asked to show this in the exercises; a similar proof is sketched in Example 3.8a.) Thus, by part (c) of Proposition 3.10, x is an accumulation point of the rational numbers. □

DEFINITION 3.12. *A set is closed if it contains all of its accumulation points.*

EXAMPLE 3.13.
a) The interval $I = [2, 4]$ is closed. We prove this by showing that if x is not in I, then x is not an accumulation point of I. This implies that if x is an accumulation point of I, then x is in I.

Assume x is not in I. Now either $x < 2$ or $x > 4$. If $x < 2$, then $2 - x$ is a positive number and the neighborhood of x with radius $2 - x$ is $(x - (2 - x), x + (2 - x))$ or $(2x - 2, 2)$. Since this neighborhood doesn't contain any point of $[2, 4]$, x is not an accumulation point of $[2, 4]$. A similar argument will show that if $y > 4$, then y is not an accumulation point of $[2, 4]$. Consequently, $[2, 4]$ must contain its accumulation points.

b) The set of rational numbers is not closed since we demonstrated a sequence of rational numbers that converged to an irrational number in Example 3.11b. □

We have seen that the rational numbers are neither open nor closed in the real numbers, so it is not true that all sets must be either open or closed. In the exercises, we ask the reader to show that the interval $[1, 2)$ is neither open nor closed.

If A is a set of real numbers, then the *complement* of A is the set of all numbers *not* contained in A. For example, the complement of $[1, 2)$ is $(-\infty, 1) \cup [2, \infty)$. Often we can discover properties of a set by examining its complement. The following proposition employs this technique.

PROPOSITION 3.14. *A set is open if and only if its complement is closed.*

PROOF. Let A be a subset of the real numbers and let B be the complement of A. We begin by showing that if A is open, then B is closed.

Suppose that A is open. If x is an accumulation point of B, then every neighborhood of x contains an element of B. Consequently, there is no neighborhood of x that is completely contained in A. As A is open, Proposition 3.3 implies that x is not in A. It follows that x is in B and so B contains all of its accumulation points. Hence, if A is open, then B is closed.

It remains to show that B is open when A is closed.

Suppose that A is closed. Let x be a real number such that for all $\epsilon > 0$ there exists y in A satisfying $|x - y| < \epsilon$. Then x must be in A, and it follows that B is open. Why? □

A number of observations follow from Proposition 3.14. For example, we now know the set of irrational numbers is neither open nor closed. This is a consequence of Proposition 3.14 and our earlier observation that the set of rational numbers is neither open nor closed. Additional implications of Proposition 3.14 appear in the exercises and later in the text.

We occasionally find it useful to speak of a collection of sets such as $\{A_1, A_2, A_3, \dots\}$, which we denote as $\{A_n\}$. We denote the union of this collection, $A_1 \cup A_2 \cup A_3 \cup \dots$, as $\bigcup A_n$ and the intersection of this collection, $A_1 \cap A_2 \cap A_3 \cap \dots$, as $\bigcap A_n$. The following proposition outlines a few of the properties of collections of open and closed sets.

PROPOSITION 3.15. *Let $\{A_n\}$ be a collection of open sets. Then $\bigcup A_n$ is open. If the collection contains only a finite number of sets, then $\bigcap A_n$ is open.*

Let $\{B_n\}$ be a collection of closed sets. Then $\bigcap B_n$ is closed. If the collection contains only a finite number of sets, then $\bigcup B_n$ is closed.

It is important to note that the intersection of an infinite number of open sets might not be open. Likewise, the union of an infinite number of closed sets might not be closed. For example, if we define the infinite collection of open sets $\{A_n\}$ by $A_n = (-\frac{1}{n}, \frac{1}{n})$, then $\bigcap A_n = \{0\}$, which is not open! Similarly, if we define the infinite collection of closed sets $\{B_n\}$ by $B_n = [-1 + \frac{1}{n}, 1 - \frac{1}{n}]$, then the union $\bigcup B_n$ is equal to $(-1, 1)$, an open set. This illustrates that the distinction between infinite and finite collections is not made lightly. A proof of Proposition 3.15 can be worked out by the reader or found in W. Rudin's text, *Principles of Mathematical Analysis*.

The last topological concept we introduce in this chapter is the concept of a dense subset.

DEFINITION 3.16. *Let A be a subset of B. Then A is dense in B if every point in B is an accumulation point of A, a point of A, or both.*

We can think of a dense subset of B as having parts in every nook and cranny of B. Another way of defining a dense subset is to say that A is dense in B if every circle centered at a point of B contains a point of A, no matter how small the radius of the circle. This idea is made precise in part (b) of the following proposition.

PROPOSITION 3.17. *Let A be a subset of B. Then, the following statements are equivalent:*

a) *A is dense in B.*

b) *For each point x in B and each $\epsilon > 0$, there exists y in A such that $|x - y| < \epsilon$.*

c) *For every point x in B, there exists a sequence of points contained in A that converges to x.*

The proof of this proposition is not hard and is left as an exercise.

EXAMPLE 3.18.

The rationals are dense in $\mathbb{R}$. To demonstrate this, it suffices to show that every irrational number is an accumulation point of the rationals.

Let $t = **.t_1 t_2 t_3 t_4 \ldots$ be an irrational number expressed in decimal notation. Define a sequence of rational numbers by

$$x_1 = **.t_1$$
$$x_2 = **.t_1 t_2$$
$$x_3 = **.t_1 t_2 t_3$$
$$x_4 = **.t_1 t_2 t_3 t_4$$
$$\vdots$$

It is easy to prove that this sequence converges to t; see Example 3.8a for a similar proof. Hence, the rationals are dense in the real numbers. □

It is interesting to note that the irrational numbers are also dense in the real numbers. This demonstrates that even though a set is large enough to be dense, it doesn't necessarily have to be so large that there isn't enough room left over for other large (that is, dense) sets. In fact, we can find an infinite collection of subsets of the real numbers, each of which is dense in the real numbers, such that no two of the sets have a point in common. Next time you are waiting patiently in a doctor's office or for a lecture to end, you might try to do just that.

Exercise Set 3

•3.1 Show that $|x - y| < \epsilon$ if and only if y is in the open interval $(x-\epsilon, x+\epsilon)$. Thus, the ϵ-neighborhood of x or $N_\epsilon(x)$ is the interval $(x - \epsilon, x + \epsilon)$.

3.2 Prove that the interval (a, b) is open. (Note that most of the proof is contained in Example 3.5.)

3.3 Prove that the interval $[a, b]$ is closed.

3.4 Show that the interval $[1,2)$ is neither open nor closed.

3.5 Prove that the rationals are not an open set.

3.6 Indicate whether the following sets are open, closed, neither open nor closed, or both open and closed. Justify your answer.

   a) $[-1, 0) \cup (0, 1]$

   b) $(-1, 0] \cup [0, 1)$

   c) $(-1, 0] \cap [0, 1)$

   d) The set of natural numbers, $\{1, 2, 3, \ldots\}$

   e) $\bigcup_{n=1}^{\infty} (\frac{-1}{n}, \frac{1}{n})$

   f) $\bigcap_{n=1}^{\infty} (0, \frac{1}{n})$

3.7 Prove Propositions 3.3, 3.4, and 3.10.

*3.8 Prove Theorem 3.6.
   Hint: Use Lemma 2.10.

3.9 Explain the second half of the proof of Proposition 3.14.

3.10 Prove that the sequence $x, x, x, \ldots$ converges to x.

3.11 Let $x = .123456789101112\ldots$ and define a sequence by

$$x_1 = .1$$
$$x_2 = .12$$
$$x_3 = .123$$
$$x_4 = .1234$$
$$\vdots$$

Show that the sequence converges to x.
Hint: Consider Example 3.8a.

3.12 Prove that $\mathbb{R}$ is both open and closed. Use Proposition 3.14 to show the empty set $\emptyset$ is both open and closed. Are there any other subsets of $\mathbb{R}$ that are both open and closed?

3.13 Show that the set $\{\frac{1}{2}, \frac{1}{3}, \frac{1}{4}, \frac{1}{5}, \ldots, 0\}$ is closed.

3.14 Let A be a closed and dense subset of B. Prove that $A = B$.

3.15 Show that the set of irrational numbers is dense in $\mathbb{R}$.

*3.16 Prove Proposition 3.15.

3.17 Prove Proposition 3.17.

4

Periodic Points and Stable Sets

We are now ready to examine the dynamics of discrete systems. We begin by defining and categorizing the simplest types of behavior. In most of what follows, we will assume that the range of the function in question is a subset of the domain. Exceptions to this practice will be noted.

DEFINITION 4.1. *If f is a function and $f(c) = c$, then c is a fixed point of f.*

A function of the real numbers has a fixed point at c if and only if the point (c, c) is on its graph. Thus, a function has a fixed point at c if and only if its graph intersects the line $y = x$ at the point (c, c).

Fixed points are important in dynamics. In Chapter 1, we observed that the function $f(x) = -x^3$ has a fixed point at 0 that attracts all of the points in the interval $(-1, 1)$ to it under iteration of the function. We will see that in many dynamical systems fixed points play a similar role.

The following theorem will help us to locate fixed points.

THEOREM 4.2. *Let $I = [a, b]$ be a closed interval and $f : I \to I$ be a continuous function. Then f has a fixed point in I.*

PROOF Let $I = [a, b]$ and $f : I \to I$ be continuous. If $f(a) = a$ or $f(b) = b$, then either a or b is fixed and we are done. Suppose $f(a) \neq a$ and $f(b) \neq b$. Let $g(x) = f(x) - x$. Since $g(x)$ is the difference of continuous functions, it is a continuous function. As $f(a) \neq a$ and $f(a)$ is in $[a, b]$, $f(a) > a$. Likewise, $f(b) < b$. Hence, $g(a) = f(a) - a > 0$ and

32   4. Periodic Points and Stable Sets

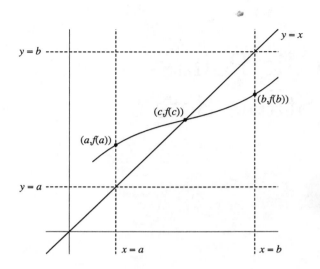

FIGURE 4.1. An illustration of the situation in Theorem 4.2. Recall that f has a fixed point at c if and only if the graph of f intersects the line $y = x$ at c. We see that such an intersection is required by the conditions on f.

$g(b) = f(b) - b < 0$. Since g is continuous, the Intermediate Value Theorem implies that there is c in $[a, b]$ such that $g(c) = 0$. But $g(c) = f(c) - c = 0$ so that $f(c) = c$, and we are done. Figure 4.1 illustrates this proof. □

EXAMPLE 4.3.
The function $f(x) = 1 - x^2$ has a fixed point in the interval $[0,1]$.

To see this, we might note that f is a continuous function and the range of f over $[0, 1]$ is contained in $[0, 1]$. Thus, by Theorem 4.2, f has a fixed point in $[0, 1]$.

Alternatively, we look at the graph of f shown in Figure 4.2 and note that it crosses the line $y = x$ in the interval $[0, 1]$. Hence, f has a fixed point in $[0, 1]$.

Finally, we can find the fixed point algebraically by solving the equation $f(x) = x$ or $1 - x^2 = x$. By doing so, we find that the fixed points of f are at $-\frac{1}{2} - \frac{\sqrt{5}}{2}$ and $-\frac{1}{2} + \frac{\sqrt{5}}{2}$. Note that $-\frac{1}{2} + \frac{\sqrt{5}}{2}$ is in $[0, 1]$. □

In Theorem 4.2, we proved that if the image of a closed interval under a continuous map is contained in the interval, then the map has a fixed point

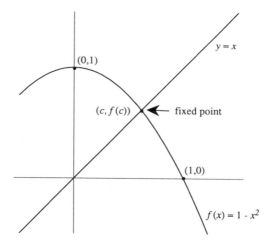

FIGURE 4.2. The graph of $f(x) = 1 - x^2$.

in the interval. In Theorem 4.4, we will show that there is a fixed point in the interval when the containment is reversed.

THEOREM 4.4. *Let I be a closed interval and $f : I \to \mathbb{R}$ be a continuous function. If $f(I) \supset I$, then f has a fixed point in I.*

PROOF. Let $I = [a, b]$. Since $f(I) \supset I$ there are c and d in I such that $f(c) = a$ and $f(d) = b$. If $c = a$ or $d = b$, we are done. If not, then $a < c < b$ and $a < d < b$. If we define $g(x) = f(x) - x$, then $g(c) = f(c) - c < 0$ since $f(c) = a$ and $a < c$. Likewise, $g(d) = f(d) - d > 0$. Since $g(c) < 0$ and $g(d) > 0$, and g is continuous, the Intermediate Value Theorem implies that there is e between c and d (and hence in I) satisfying $g(e) = 0$ and $f(e) = e$, which complete the proof. The reader is encouraged to draw a figure that illustrates this argument. □

We noted earlier that the function $f(x) = -x^3$ examined in Chapter 1 has a fixed point at 0. The behavior of the points 1 and -1 under iteration of f is also worth noting. Recall, f maps 1 to -1 and -1 back to 1. Consequently, we call 1 and -1 periodic points and the set $\{-1, 1\}$ a periodic orbit.

DEFINITION 4.5. *Let f be a function. The point x is a periodic point of f with period k if $f^k(x) = x$. In other words, a point is a periodic point of f with period k if it is a fixed point of f^k. The periodic point x has prime period k_0 if $f^{k_0}(x) = x$ and $f^n(x) \neq x$ whenever $0 < n < k_0$. That* is, a periodic point has prime period k_0 if it returns to its starting place for the first time after exactly k_0 iterations of f.

The set of all iterates of the point x is called the orbit of x, and if x is a periodic point, then it and its iterates are called a periodic orbit or a periodic cycle.

To illustrate Definition 4.5, we again consider $f(x) = -x^3$. The points 1 and -1 form a periodic orbit with prime period 2. The set of points with period 2 is $\{-1, 1, 0\}$, while the set of points with prime period 2 is $\{-1, 1\}$. The only fixed point of f is 0. The orbit of 0 is $\{0\}$ and the orbit of 2 is $\{2, -2^3, 2^9, -2^{27}, \dots\}$.

A function may have many fixed or periodic points. The function $f(x) = x$ fixes every point, while every point except 0 is a periodic point with prime period 2 for $g(x) = -x$. Later, we shall see that, for some values of r, the function $h(x) = rx(1-x)$ has a periodic point with each prime period.

Two other types of points are eventually fixed points and eventually periodic points. For example, the function $h(x) = 4x(1-x)$ has a fixed point at 0. The point 1 is not fixed, but $h(1) = 0$, so after one iteration 1 is fixed at 0. Also, $h(\frac{1}{2}) = 1$ and $h^2(\frac{1}{2}) = 0$ so after two iterations $\frac{1}{2}$ is fixed at 0. In the exercises, we ask the reader to find two points that are fixed at 0 after exactly three iterations of h. Given sufficient patience, it is possible to find points that are fixed by h after any number of iterations.

Before we look at an example of a point that is eventually periodic but not eventually fixed, we introduce a precise definition of these terms.

DEFINITION 4.6. *Let f be a function. The point x is an eventually fixed point of f if there exists N such that $f^{n+1}(x) = f^n(x)$ whenever $n \geq N$. The point x is eventually periodic with period k if there exists N such that $f^{n+k}(x) = f^n(x)$ whenever $n \geq N$.*

EXAMPLE 4.7.
Let $g(x) = |x - 1|$. Then 0 and 1 form a periodic cycle and $f(2) = 1$, so 2 is eventually periodic. In fact, every integer is eventually periodic. □

The function $f(x) = -x^3$ has a fixed point at 0 and a periodic cycle consisting of 1 and -1. Examination reveals that there are no eventually periodic points that are not periodic. How should we characterize the rest of the points? For example, if we start at $\frac{1}{2}$ and iterate with f, then we get the sequence, $\frac{1}{2}, -\frac{1}{2^3}, \frac{1}{2^9}, -\frac{1}{2^{27}}, \dots$. So $\frac{1}{2}$ is not periodic and never reaches 0, though it does get closer and closer to it. That is, $f^n(\frac{1}{2})$ converges to 0 as n goes to ∞. We say $\frac{1}{2}$ is forward asymptotic to 0.

DEFINITION 4.8. *Let f be a function and p be a periodic point of f with period k. Then x is forward asymptotic to p if the sequence x, $f^k(x)$, $f^{2k}(x)$, $f^{3k}(x)$, ... converges to p. In other words, $\lim_{n\to\infty} f^{nk}(x) = p$. The stable set of p, denoted by $W^s(p)$, consists of all points that are forward asymptotic to p.*

If the sequence $|x|$, $|f(x)|$, $|f^2(x)|$, $|f^3(x)|$, ... grows without bound, then x is forward asymptotic to infinity. The stable set of infinity, denoted by $W^s(\infty)$, consists of all points that are forward asymptotic to infinity.

Notice that when we are searching for points in the stable set of a point with prime period k, we must consider the sequence

$$x, f^k(x), f^{2k}(x), f^{3k}(x), \ldots,$$

not the sequence

$$x, f(x), f^2(x), f^3(x), \ldots.$$

EXAMPLE 4.9.

a) Let $f(x) = -x^3$. Then the stable set of 0 consists of all points in the interval $(-1, 1)$, the stable set of infinity consists of all points in the intervals $(-\infty, -1)$ and $(1, \infty)$, the stable set of -1 contains only the point -1, and the stable set of 1 contains only the point 1. Symbolically, we write $W^s(0) = (-1, 1)$, $W^s(\infty) = (-\infty, -1) \cup (1, \infty)$, $W^s(-1) = \{-1\}$, and $W^s(1) = \{1\}$. Notice that it doesn't make sense to discuss the stable set of any other points since f has no other periodic points.

b) In Example 4.7, we examined the function $g(x) = |x - 1|$. Returning to that example, we see $W^s(0)$ is the set of all even integers and $W^s(1)$ is the set of all odd integers. □

In the preceding example, we tacitly assumed that if a point is in the stable set of one periodic point, then it cannot be in the stable set of a different periodic point. This idea is important, and we prove it in the following theorem.

THEOREM 4.10. *The stable sets of distinct periodic points do not intersect. In other words, if p_1 and p_2 are periodic points and $p_1 \neq p_2$, then $W^s(p_1) \cap W^s(p_2) = \emptyset$.*

PROOF. Let $f(x)$ be a function with periodic points p_1 and p_2 of period k_1 and k_2, respectively. We'll show that if $W^s(p_1) \cap W^s(p_2) \neq \emptyset$, then $p_1 = p_2$.

Let x be in $W^s(p_1) \cap W^s(p_2)$. Then for each $\epsilon > 0$ there exist N_1 and N_2 such that $n \geq N_1$ implies $|p_1 - f^{nk_1}(x)| < \frac{\epsilon}{2}$ and $n \geq N_2$ implies $|p_2 - f^{nk_2}(x)| < \frac{\epsilon}{2}$. If M is the larger of N_1 and N_2, then $n \geq M$ implies both $|p_1 - f^{nk_1}(x)| < \frac{\epsilon}{2}$ and $|p_2 - f^{nk_2}(x)| < \frac{\epsilon}{2}$. Utilizing the Triangle Inequality, we find that when $n \geq M$, then

$$\begin{aligned}|p_1 - p_2| &= |p_1 - f^{nk_1k_2}(x) + f^{nk_1k_2}(x) - p_2| \\ &\leq |p_1 - f^{(nk_2)k_1}(x)| + |f^{(nk_1)k_2}(x) - p_2| \\ &< \frac{\epsilon}{2} + \frac{\epsilon}{2} = \epsilon.\end{aligned}$$

Since we have shown that the distance between p_1 and p_2 is less than ϵ for each $\epsilon > 0$, it must be that $p_1 = p_2$. □

Now that we have defined periodic points, eventually periodic points, and stable sets, we are able to classify most points in a simple dynamical system. In our first example, $f(x) = -x^3$, we have three types of points: the fixed point 0, the periodic points 1 and -1, and points in the stable sets of 0 or ∞. As we continue our investigations, we shall find that not all systems are quite so easy to characterize. We will need new tools and ideas to classify the dynamics of even a simple function like $h(x) = 4x(1-x)$. One of the simplest tools we have at our disposal is graphical analysis.

4.1. Graphical Analysis

As the name implies, graphical analysis uses the graph of a function to analyze its dynamics. It is best understood by studying a few examples.

EXAMPLE 4.11.
We begin our examination of the dynamics of the function $f(x) = x^3$ by graphing f and the line $y = x$ on the same set of coordinate axes as in Figure 4.3. We then try to determine the itinerary of the point a that lies in the interval (0,1). Rather than follow the movement of the point on the x-axis, we keep track of its whereabouts on the line $y = x$. It starts at a (or, if you prefer, at the point (a, a)). From that point, we travel vertically until we strike the graph of f. Since we moved vertically, the x value of the point of intersection must be a and the y value must be $f(a)$. We now travel horizontally back to the line $y = x$ to arrive at the point $f(a)$ (or $(f(a), f(a))$). Repeating the process, we travel vertically to the graph of f to arrive at the point $(f(a), f^2(a))$ and then horizontally back to $f^2(a)$ on the $y = x$ line. Continuing, we see that $f^n(a)$ approaches 0 as n goes to infinity. After trying more points in the interval $(0, 1)$, we conclude that every point in $(0, 1)$ will approach 0 under iteration of $f(x) = x^3$. Further examination suggests that $W^s(0) = (-1, 1)$ and $W^s(\infty) = (-\infty, -1) \cup (1, \infty)$. The points -1, 0, and 1 are fixed points of f. □

4.1. Graphical Analysis   37

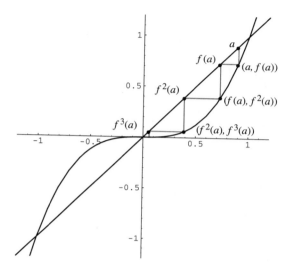

FIGURE 4.3. Graphical analysis of the orbit of the point a under iteration of $f(x) = x^3$.

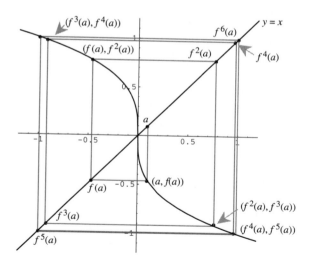

FIGURE 4.4. Graphical analysis of the orbit of a under iteration of $g(x) = -x^{1/3}$.

Example 4.12.

We now consider the function $g(x) = -x^{1/3}$. The graph of g and the line $y = x$ are shown in Figure 4.4 on the previous page. We start with a point a which is near 0 and in the interval $(0,1)$. Note that as n grows, the value of $g^n(a)$ approaches 1 in absolute value and oscillates from one side of 0 to the other. Note also that -1 and 1 form a periodic cycle with prime period 2. Further, $g^n(a)$ is always positive when n is even. Thus, we conclude that a is in $W^s(1)$.

Further analysis suggests that $W^s(1) = (0, \infty)$, $W^s(-1) = (-\infty, 0)$, 0 is fixed, and $W^s(\infty) = \emptyset$. □

The reader is asked to develop additional examples of graphical analysis in the exercises.

Exercise Set 4

4.1 Let $g(x) = |x - 1|$. Find all fixed points and eventually fixed points of g. How many periodic points does g have? What are they? How many eventually periodic points does g have? What are they?

4.2 From Exercise 4.1 we know that $\frac{1}{2}$ is fixed by $g(x) = |x - 1|$. Find $W^s(\frac{1}{2})$.

4.3 Find two points in $[0, 1]$ that are eventually fixed at 0 after exactly three iterations of $h(x) = 4x(1 - x)$.

4.4 a) Can a function have a point that is eventually fixed after exactly two iterations if it has no points that are eventually fixed after exactly one iteration? If so, give an example; if not, explain why.

b) Suppose n is a natural number. Can a function have a point that is eventually fixed after exactly $n + 1$ iterations if it has no points that are eventually fixed after exactly n iterations? If so, give an example; if not, explain why.

4.5 Show that all periodic points of $h(x) = 3.2x(1 - x)$ lie in the interval $[0, 1]$.

4.6 In the text we stated that $f(x) = -x^3$ has no eventually fixed points (page 34). Explain why this is true.

4.7 For each of the following functions, find all periodic points and describe the stable sets of each. You may wish to use graphical analysis.

   a) $q(x) = x^2$

   b) $E(x) = e^x - 1$

   c) $p(x) = x^2 - x$

   d) $r(x) = -\frac{4}{\pi} \arctan x$

   e) $q(x) = x^2 + \frac{1}{4}$

   * f) $h(x) = 3.5x(1-x)$

4.8 Let p be a fixed point of the function $f(x)$ with prime period k.

   a) Show that $f(p), f^2(p), \ldots, f^{k-1}(p)$ are also periodic points of f with prime period k.

   b) Prove that $W^s(f(p)) = f(W^s(p))$.

4.9 Find a function $f : [0,1] \to [0,1]$ with no fixed points and draw its graph.

4.10 Find a continuous function $f : [0,1] \to [0,1]$ with exactly three fixed points and draw its graph.

4.11 Find a continuous function $f : (1,3) \to (1,3)$ that doesn't have a fixed point. Explain why this doesn't contradict Theorem 4.2.

4.12 Draw a figure that illustrates the proof of Theorem 4.4. Use your figure to explain the proof.

4.13 A function f has a fixed point at c if (c,c) is on the graph of f. Suppose (a,b) is on the graph of f and a is a period 2 point of f. What other point must be on the graph of f? Can you generalize this to a period n orbit?

4.14 Find a continuous function of the real numbers that has a periodic point of period 3 and indicate the periodic point.
   * What other periodic points does the function have?
   Hint: Choose the periodic orbit first. Use the fact that (a,b) is on the graph of f if and only if $f(a) = b$ and find a continuous function whose graph goes through the appropriate points.

5
Sarkovskii's Theorem

Consider the function $p(x) = -\frac{3}{2}x^2 + \frac{5}{2}x + 1$. It is easy to see that $p(0) = 1$, $p(1) = 2$, and $p(2) = 0$. So $\{0, 1, 2\}$ is an orbit with period three. It is reasonable to ask how many other periodic points $p(x)$ has and what prime periods are represented. These questions are answered, at least in part, by the following remarkable theorem:

THEOREM 5.1. *If a continuous function of the real numbers has a periodic point with prime period three, then it has a periodic point of each prime period. That is, for each natural number n there is a periodic point with prime period n.*

PROOF. Let $\{a, b, c\}$ be a period three orbit of the continuous function f. Without loss of generality, we assume $a < b < c$. There are two cases: $f(a) = b$ or $f(a) = c$. We suppose $f(a) = b$. This implies $f(b) = c$ and $f(c) = a$. The proof of the case $f(a) = c$ is similar.

Let $I_0 = [a, b]$ and $I_1 = [b, c]$. The Intermediate Value Theorem implies that $f(I_0) \supset I_1$, $f(I_1) \supset I_1$, and $f(I_1) \supset I_0$. (See Exercise 2.16.) Since $f(I_1) \supset I_1$, Theorem 4.4 implies that f has a fixed point in I_1, that is, f has a periodic point with prime period 1.

Now let n be a natural number larger than 1. We want to show that f has a periodic point with prime period n. Since a has prime period 3, the case $n = 3$ is done and we may assume that $n \neq 3$. To find an appropriate periodic point for n, we use a nested sequence of closed intervals, $I_1 = A_0 \supset A_1 \supset A_2 \supset \cdots \supset A_n$, with the following properties:

(1) $A_0 = I_1$
(2) $f(A_k) = A_{k-1}$ for $k = 1, 2, \ldots, n-2$
(3) $f^k(A_k) = I_1$ for $k = 1, 2, \ldots, n-2$
(4) $f^{n-1}(A_{n-1}) = I_0$
(5) $f^n(A_n) = I_1$

We proceed by first showing that the existence of such a sequence of sets implies that there is a point in A_n with prime period n. We then show that such a sequence exists.

Since $A_n \subset I_1$, property (5) and Theorem 4.4 imply that f^n has a fixed point in A_n. Of course, this is equivalent to saying that f has a periodic point p with period n in A_n. We use the other four properties to show that p has prime period n.

Let p be a periodic point in A_n with period n. Since A_n is in I_1, we know that p is in $I_1 = [b, c]$. We also note that property (3) implies that the points $f(p), f^2(p), \ldots, f^{n-2}(p)$ are in $I_1 = [b, c]$ and property (4) implies that $f^{n-1}(p)$ is in $I_0 = [a, b]$. We now show that p is not b or c by contradiction. Suppose that $p = c$. Then $f(p) = a$, which is not in I_1. Since $f^{n-1}(p)$ is the only iterate of the first n iterates of p that isn't in I_1, it must be that $n = 2$. But this contradicts the fact that the prime period of c is three, and so $p \neq c$. To see that $p \neq b$, we note that if $p = b$, then $n = 3$ since $f^2(p) = a$, which is not in I_1, and the only iterate of the first n iterates of p that is not in I_1 is $f^{n-1}(p)$. Since we have assumed that $n \neq 3$, we conclude that $p \neq b$. The preceding arguments demonstrate that p is not b or c and so p must be in the open interval (b, c).

Since $f^{n-1}(p)$ is in $I_0 = [a, b]$, which is disjoint from (b, c), $f^{n-1}(p)$ is not equal to p and so p can't have prime period $n-1$. If the prime period of p were less than $n-1$, then property (3) and the fact that p is not b or c would imply that the orbit of p is contained entirely in (b, c), and this would contradict property (4). So, p must have prime period n. Therefore, if a sequence of closed sets with the required properties exists for n, then there is a point p with prime period n.

To complete the proof of this theorem, we demonstrate that such a sequence of closed sets exists for each natural number larger than 1. In doing this, we will use the following lemma.

LEMMA 5.2. *Let $J = [a, b]$ and $I = [c, d]$ be closed intervals and let f be a continuous function satisfying $f(J) \supset I$. Then there exists an interval J_0 such that $J_0 \subset J$ and $f(J_0) = I$.*

This is an intuitively clear result the proof of which depends on the continuity of f. It is easy to convince oneself of the truth of the lemma by drawing suitable graphs. In particular, one quickly finds that it is impossible to draw the graph of a continuous function for which the lemma wouldn't hold. An analytic proof is outlined in Exercise 5.7 at the end of this chapter.

Now, to complete the proof of the theorem, let n be a natural number larger than 1. We create the desired sequence and establish each of the requirements in turn. Obviously, we can choose A_0 so that $A_0 = I_1$ and property (1) is satisfied. Since $A_0 = I_1$ and $f(I_1) \supset I_1$, we have $f(A_0) \supset A_0$, and so Lemma 5.2 implies there is A_1 contained in A_0 such that $f(A_1) = A_0$. Then $A_1 \subset A_0$ implies that $f(A_1) \supset A_1$. Consequently, by Lemma 5.2 there is $A_2 \subset A_1$ such that $f(A_2) = A_1$. We continue in this manner and define A_k for $k = 1, 2, \ldots, n-2$. In each case, we find A_k contained in A_{k-1} so that $f(A_k) = A_{k-1}$ for $k = 1, 2, \ldots, n-2$ as required by property (2). Notice that $f(A_k) \supset A_k$ for each k, so Lemma 5.2 implies that the process of defining the intervals A_k can continue indefinitely. To demonstrate that property (3) holds, we note that property (2) implies

$$f^2(A_k) = f(f(A_k)) = f(A_{k-1}) = A_{k-2}$$
$$f^3(A_k) = f(f^2(A_k)) = f(A_{k-2}) = A_{k-3}$$
$$\vdots$$
$$f^{k-1}(A_k) = f(f^{k-2}(A_k)) = f(A_{k-(k-2)}) = f(A_2) = A_1$$
$$f^k(A_k) = f(f^{k-1}(A_k)) = f(A_1) = A_0 = I_1$$

for each $k = 1, 2, \ldots, n-2$, as required by (3). To define A_{n-1} consistent with property (4), we note that

$$f^{n-1}(A_{n-2}) = f(f^{n-2}(A_{n-2})) = f(I_1).$$

Since $f(I_1) \supset I_0$, we know that $f^{n-1}(A_{n-2}) \supset I_0$. Hence, by Lemma 5.2 there is $A_{n-1} \subset A_{n-2}$ such that $f^{n-1}(A_{n-1}) = I_0$. Finally,

$$f^n(A_{n-1}) = f(f^{n-1}(A_{n-1})) = f(I_0)$$

and $f(I_0) \supset I_1$ imply that $f^n(A_{n-1}) \supset I_1$. Again using Lemma 5.2, we see that there is $A_n \subset A_{n-1}$ such that $f^n(A_n) = I_1$ as required by property (5), and the proof is complete. □

As surprising as Theorem 5.1 might be, it is only a special case of a more general theorem proven by A. N. Sarkovskii in 1964. This interesting and beautiful result depends only on the continuity of the function in question and should give pause to anyone who thinks that all of the really good theorems using only elementary properties of functions were proven before the dawn of the twentieth century.

In his theorem, Sarkovskii places an order on the integers that we define here. The statement of the theorem follows the definition.

DEFINITION 5.3. SARKOVSKII'S ORDERING. *Sarkovskii's ordering of the natural numbers is*

$$3 \succ 5 \succ 7 \succ \cdots \succ 2 \cdot 3 \succ 2 \cdot 5 \succ 2 \cdot 7 \succ \cdots$$
$$\cdots \succ 2^2 \cdot 3 \succ 2^2 \cdot 5 \succ \succ 2^2 \cdot 7 \succ \cdots \succ 2^n \cdot 3 \succ 2^n \cdot 5 \succ 2^n \cdot 7 \succ \cdots$$
$$\cdots \succ 2^3 \succ 2^2 \succ 2 \succ 1.$$

The relation $a \succ b$ indicates a precedes b in the order. When writing the order, all odd numbers except one are listed in ascending order, then two times every odd, then four times every odd, and so on. The order is completed by listing the powers of two in descending order. Every natural number can be found exactly once in Sarkovskii's ordering.

THEOREM 5.4. [SARKOVSKII, 1964] *Suppose that f is continuous function of the real numbers and that f has a periodic point with prime period n. If $n \succ m$ in Sarkovskii's ordering, then f also has a periodic point with prime period m.*

While the proof of Sarkovskii's Theorem uses no tools that are not used in the proof of Theorem 5.1, it is somewhat longer. In spirit it is much the same. Essentially, we show that if $n \succ m$, then the existence of a prime period n point implies the existence of a sequence of closed intervals $I_1, I_2, \ldots, I_m$, such that $f(I_i) \supset I_{i+1}$ for all $i < m$ and $f(I_m) \supset I_1$. Then Theorem 4.4 implies there is a period m point in I_1. If one of the intervals is disjoint from all of the other intervals, then there is a point with prime period m. The key then is to find these intervals I_i. Rather than write the proof out in detail here, we refer the reader to the proof by Huang on pages 91–102 of the April 1992 issue of *Mathematics Magazine*. Huang's exposition is clear and complete. A similar proof can be found in Devaney's text, *Introduction to Chaotic Dynamics,* which is listed in the references. Both authors demonstrate that Sarkovskii's Theorem is sharp by constructing continuous functions that have period five points, but no period three points. The technique used allows for the construction of a continuous function that has a periodic point with prime period n and no periodic point with prime period that precedes n in the Sarkovskii ordering.

In addition to implying the existence of orbits with prime periods lower than that of a known orbit, Sarkovskii's Theorem can be used to show that points with certain periods do not exist. For example, in Figure 5.1 the graphs of $h(x) = 3.2x(1-x)$, $h^2(x)$ and $h^4(x)$ are shown.

In the first graph, we see that h has fixed points at 0 and near .7. In the second graph, we see the fixed points again, and a period two orbit near .5 and .8. In the third graph, we see that the only period four points are at 0 and near .7, .5, and .8. That is, h has no prime period four orbits. Therefore, we can conclude that h has no orbits with prime period

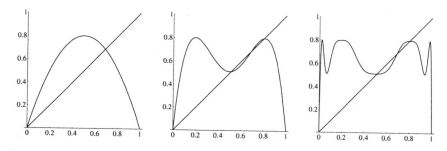

FIGURE 5.1. The graphs of $h(x) = 3.2x(1-x)$, $h^2(x)$, and $h^4(x)$.

other than 2 and 1, since those are the only two numbers less than 4 in the Sarkovskii ordering. We should note that, as was demonstrated in Exercise 4.5, all of the periodic points of h lie in the interval $[0, 1]$, so we do not need to graph any other portion of the function.

Exercise Set 5

5.1 Let I_1 and I_0 be as described in the proof of Theorem 5.1. Use the Intermediate Value Theorem to show that $f(I_0) \supset I_1$, $f(I_1) \supset I_1$, and $f(I_1) \supset I_0$.

5.2 Demonstrate the existence of an orbit of $h(x) = 4x(1-x)$ with prime period three. What does Sarkovskii's Theorem tell us about the periodic points of $h(x)$?

Hint: Use a computer graphics package to look at the graphs of h and h^3.

5.3 Show that $f(x) = \frac{-4}{\pi} \arctan(x)$ does not have a point with prime period 32. What are the possible prime periods of periodic points of f?

Hint: Use a computer graphics package to look at the graphs of the iterates of f.

5.4 Suppose that $I_0, I_1, I_2, \ldots, I_{n-1}$ are closed intervals and f is a continuous function such that $f(I_k) \supset I_{k+1}$ for $0 \leq k < n - 1$.

a) Show that I_0 contains a sequence of closed intervals $A_0 \supset A_1 \supset A_2 \supset \cdots \supset A_{n-1}$ such that $f^k(A_k) = I_k$ for $1 \leq k \leq n - 1$.

b) Show that there is a point x_0 in I_0 such that $f^i(x_0)$ is in I_i for $0 \leq i \leq n-1$.

c) Prove that if $f(I_{n-1}) \supset I_0$, then f has a periodic point in I_0 with period n.

d) Prove that if $f(I_{n-1}) \supset I_0$ and I_0 is disjoint from the other intervals, then f has a point with prime period n in I_0.

5.5 Let $f : \mathbb{R} \to \mathbb{R}$ be continuous, $n > 3$, and $x_1, x_2, \ldots, x_n$ be points such that $x_1 < x_2 < \cdots < x_n$. Show that if $f(x_i) = x_{i+1}$ for $i = 1, 2, \ldots, n-1$ and $f(x_n) = x_1$, then f has points with all prime periods.

5.6 a) Let $f : I \to I$ be a continuous function of the interval I and suppose f has a periodic point with prime period three in I. Show that f has periodic points of all orders in I.
Hint: Extend f to a continuous function of the real line.

b) Let I be an interval. State and prove an analog of Sarkovskii's Theorem for $f : I \to I$. You may assume Sarkovskii's theorem.

∗5.7 Prove Lemma 5.2.
Hint: Let x_0 be a point in $[a, b]$ such that $f(x_0) = c$. Without loss of generality, we can assume there is y in $[x_0, b]$ such that $f(y) = d$. (Explain why we can do this.) Let x_2 be the greatest lower bound of the set $\{y \text{ in } [x_0, b] \mid f(y) = d\}$ and x_1 be the least upper bound of the set $\{x \text{ in } [x_0, x_2] \mid f(x) = c\}$. Show that $f([x_1, x_2]) = [c, d]$.

6
Differentiability and Its Implications

In many cases, the functions we consider are differentiable. In this chapter we examine the dynamical information contained in the derivative of the function.

DEFINITION 6.1. *Let I be an interval, $f : I \to \mathbb{R}$, and let a be a point in I. The function is differentiable at a if the limit*

$$\lim_{x \to a} \frac{f(x) - f(a)}{x - a}$$

exists. In this case, we say f is differentiable at a and denote the limit as $f'(a)$ or the derivative of f at a. A function is differentiable if it is differentiable at each point in its domain.

Recall from calculus that a function that is differentiable at a point is also continuous at that point. An important property of differentiable functions, which we will often exploit, is the Mean Value Theorem.

THEOREM 6.2. MEAN VALUE THEOREM. *Let $f : [a,b] \to \mathbb{R}$ be differentiable on $[a,b]$. Then there exists c in $[a,b]$ such that*

$$f(b) - f(a) = f'(c)(b - a).$$

After examining a few graphs, the reader will realize that the Mean Value Theorem is an intuitively clear result. A proof can be found in any good calculus text. We use the Mean Value Theorem to get another result about fixed points. Theorem 4.2 states that every continuous function that maps a closed interval back into itself has at least one fixed point. If the function is differentiable on the closed interval and the value of the derivative is less than one in absolute value, we get a much stronger result.

THEOREM 6.3. *Let I be a closed interval and $f : I \to \mathbb{R}$ be a differentiable function satisfying $I \subset f(I)$ and $|f'(x)| < 1$ for all x in I. Then f has a unique fixed point in I. Moreover, if x and y are any two points in I and $x \neq y$, then $|f(x) - f(y)| < |x - y|$.*

In the hypothesis of Theorem 6.3, we assume f is differentiable on I. Consequently, f must also be continuous on I, and Theorem 4.4 guarantees that f has a fixed point. The additional differentiability condition guarantees that f has only one fixed point, and points in I become closer together as we iterate the function. In fact, it can be shown that I is contained in the stable set of the unique fixed point. If we assume f' is continuous, then this fact is a corollary of the proof of Theorem 6.5. If f' is not continuous, then the proof requires slightly more sophisticated tools from mathematical analysis than we have at our disposal. In particular, an understanding of least upper bounds is useful. Finally, we note that, as with Theorems 4.2 and 4.4, Theorem 6.3 does not necessarily hold when I is an open interval. (See Exercise 4.11.)

The proof of Theorem 6.3 follows from the Mean Value Theorem.

PROOF. We begin by proving the second part of the theorem. Let x and y be any two points in I satisfying $x \neq y$. Without loss of generality, we assume $x < y$. Then, our hypotheses assert that f is differentiable on the interval $[x, y]$. By the Mean Value Theorem, there exists c in $[x, y]$ such that

$$|f(x) - f(y)| = |f'(c)||x - y| \quad \text{or} \quad |f'(c)| = \frac{|f(x) - f(y)|}{|x - y|}. \quad (6.1)$$

Since c is contained in $[x, y]$ and $[x, y]$ is contained in I, we know that $|f'(c)| < 1$. Hence, equation (6.1) implies $|f(x) - f(y)| < |x - y|$, as desired.

It remains to show that f has a unique fixed point. As we noted earlier, Theorem 4.4 implies that f must have at least one fixed point in I. Let p be a fixed point and let x be any other point in I. We have already shown that

$$|p - f(x)| = |f(p) - f(x)| < |p - x|.$$

It follows that $f(x) \neq x$ and x is not a fixed point of f. Therefore, p is the only fixed point of f in I. $\square$

Notice that by assuming that the function is differentiable we are able to make a much more specific statement about the dynamics of the function. If the function is differentiable and the derivative is also continuous, we gain further information about the dynamics. To illustrate this, we consider a few simple examples.

EXAMPLE 6.4.
a) Let $f(x) = \frac{1}{2}x + \frac{3}{2}$. Solving $\frac{1}{2}x + \frac{3}{2} = x$ we find that 3 is the only fixed point of f. Notice also that $f'(x) = \frac{1}{2}$ for all x. Hence, if x is any real number other than 3, then the Mean Value Theorem implies that there exists a real number, c_1, such that

$$|f(x) - f(3)| = |f'(c_1)||x - 3|$$
$$= \tfrac{1}{2}|x - 3|.$$

Iterating, we find that there exist c_2 and c_3 such that

$$|f^2(x) - f^2(3)| = |f'(c_2)||f(x) - f(3)|$$
$$= \tfrac{1}{2}\left(\tfrac{1}{2}|x - 3|\right) = \tfrac{1}{2^2}|x - 3|$$

and

$$|f^3(x) - f^3(3)| = |f'(c_3)||f^2(x) - f^2(3)|$$
$$= \tfrac{1}{2^3}|x - 3|.$$

Continuing by induction, we establish that

$$|f^n(x) - f^n(3)| = \tfrac{1}{2^n}|x - 3|.$$

Since $f^n(3) = 3$, we have $|f^n(x) - 3| = \frac{1}{2^n}|x - 3|$, and we conclude that $f^n(x)$ converges to 3 and x is in $W^s(3)$. Consequently, $W^s(3) = \mathbb{R}$.
This is illustrated using graphical analysis in Figure 6.1.

b) Now let $r(x) = -\frac{1}{2}x + \frac{9}{2}$. Once again, 3 is the unique fixed point, but this time $r'(x) = -\frac{1}{2}$. Using the Mean Value Theorem as in part (a), we can show that $W^s(3) = \mathbb{R}$. In Figure 6.1, we illustrate this using graphical analysis. Notice the relationship between the sign of the derivative and the behavior of x under iteration of the function. For $f(x) = \frac{1}{2}x + \frac{3}{2}$, the derivative is positive and x approaches 3 from one side. For $r(x) = -\frac{1}{2}x + \frac{9}{2}$, the derivative is negative and the iterates of x alternate from one side of 3 to the other. This is a pattern we will see again.

c) Let $g(x) = 2x - 3$. Again, 3 is a fixed point, but the derivative is positive and *greater* than one in absolute value. Before we go on, stop and consider the following questions: What implications does the derivative at

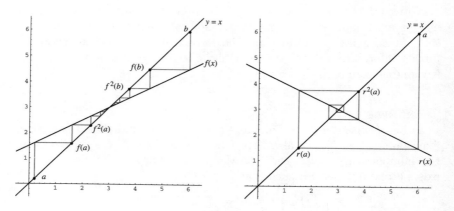

FIGURE 6.1. Graphical analysis of $f(x) = \frac{1}{2}x + \frac{3}{2}$ and $r(x) = -\frac{1}{2}x + \frac{9}{2}$.

the fixed point have for the behavior of the rest of the points? How would one prove that our hunch is correct?

Since $g'(x) = 2$ for all x, the Mean Value Theorem implies that there exist $c_1, c_2, \ldots$ such that

$$|g(x) - g(3)| = |g'(c_1)||x - 3| = 2|x - 3|$$
$$|g^2(x) - g^2(3)| = |g'(c_2)||g(x) - g(3)| = 2^2|x - 3|$$
$$\vdots$$
$$|g^n(x) - g^n(3)| = |g'(c_n)||g^{n-1}(x) - g^{n-1}(3)| = 2^n|x - 3|.$$

As $g^n(3) = 3$, we see that $|g^n(x) - 3| = 2^n|x - 3|$ and the distance from $g^n(x)$ to 3 grows without bound as n goes to infinity. Consequently, x is in $W^s(\infty)$ and $W^s(\infty) = (-\infty, 3) \cup (3, \infty)$.

d) That leaves us one more case, $k(x) = -2x + 9$. Once again, 3 is a fixed point. This time the derivative is -2, which is both negative and greater than one in absolute value. As the reader should be able to show, $W^s(\infty) = (-\infty, 3) \cup (3, \infty)$, and the iterates of x alternate from one side of 3 to the other as they grow without bound. □

Example 6.4 suggests the following theorem.

THEOREM 6.5. *Let f be a differentiable function, let p be a fixed point of f, and suppose f' is continuous. If $|f'(p)| < 1$, then there exists an open interval U containing p such that whenever x is in U, then $f^n(x)$ converges to p. That is, $U \subset W^s(p)$. If $|f'(p)| > 1$, then there exists an open interval* containing p such that all points in the interval that are not equal to p must leave the interval under iteration of f.

Before we prove Theorem 6.5, we make a few observations. First, we use the fact that f' is continuous to prove the theorem. However, this is only for our convenience; the theorem is true even if f' isn't continuous. (See the proof of Theorem 14.11 for an outline of a proof of this fact.) It is difficult to imagine a differentiable function whose derivative is not continuous, but such functions do exist. This is a point worth pondering that should be addressed in any good class on real analysis. A function whose derivative exists and is continuous is called C^1.

Second, in the case where $|f'(p)| > 1$, it is not generally true that once a point leaves the open interval containing p it is gone for good. When we examine the logistic function more carefully, we will see cases where a point's iterates return over and over again. As we demonstrated in Chapter 3, we may assume the open set containing p is a neighborhood of p.

Finally, as we noted in Example 6.4, it is true that when the derivative is negative on the open interval U, then x and $f(x)$ are on opposite sides of p, and when the derivative is positive on U, then x and $f(x)$ are on the same side of p. The reader is asked to prove this in the exercises.

To sum up, we restate Theorem 6.5.

THEOREM 6.5 (RESTATED). *Let f be a C^1 function and p be a fixed point of f. Then $|f'(p)| < 1$ implies that there is a neighborhood of p contained in $W^s(p)$ and $|f'(p)| > 1$ implies that there is a neighborhood of p in which all points other than p must leave the neighborhood under iteration of f.*

PROOF. Let f be differentiable with a continuous derivative, assume that p is a fixed point of f, and suppose that $|f'(p)| < 1$. We wish to find a neighborhood of p that is contained in $W^s(p)$. Our strategy is to find $\lambda < 1$ and $\delta > 0$ such that $|f'(x)| < \lambda$ for all x in $(p - \delta, p + \delta)$. We then apply the Mean Value Theorem to show that $(p - \delta, p + \delta) \subset W^s(p)$.

We begin by choosing $\lambda = \frac{1}{2}(1 + |f'(p)|)$ so that $|f'(p)| < \lambda < 1$. To find δ, we use the continuity of f'. Since the absolute value function and f' are continuous, the function $|f'(x)|$ is continuous. Then, Lemma 2.10 implies that there is $\delta > 0$ such that $|f'(x)| < \lambda$ for all x in $(p-\delta, p+\delta)$. It remains to show that $f^n(x)$ converges to p whenever x is in $(p - \delta, p + \delta)$.

Let x be in $(p-\delta, p+\delta)$ and not equal to p. By the Mean Value Theorem, there is c between x and p such that

$$|f(x) - p| = |f(x) - f(p)| = |f'(c)||x - p|.$$

Since c must be in the interval $(p - \delta, p + \delta)$ this implies

$$|f(x) - p| < \lambda|x - p|.$$

As the distance from $f(x)$ to p is less than the distance from x to p, $f(x)$ must be in $(p-\delta, p+\delta)$, and we can iterate the previous argument to show that
$$|f^2(x) - p| < \lambda^2 |x - p|.$$
Continuing by induction, we find
$$|f^n(x) - p| < \lambda^n |x - p|.$$
Now, since $\lambda < 1$, it follows that λ^n converges to 0 as n goes to infinity. Thus, we can conclude that $|f^n(x) - p|$ approaches 0 and $f^n(x)$ converges to p as n goes to infinity.

The proof of the case where $|f'(p)| > 1$ is similar and is left as an exercise. □

Fixed points whose derivatives are not equal to one in absolute value are important enough to have their own name. They are called *hyperbolic fixed points*. For obvious reasons, fixed points whose derivatives are less than one in absolute value are said to be *attracting*, and fixed points whose derivatives are greater than one in absolute value are said to be *repelling*. If the derivative of a fixed point is 1 or -1, it is called a *nonhyperbolic* or *neutral fixed point*.

In Theorem 6.5, we saw that we can make some predictions about the behavior of points in the neighborhood of a hyperbolic fixed point. In the following example, we see that life is not quite so predictable near nonhyperbolic fixed points. In particular, we examine three functions, each with a nonhyperbolic fixed point at 0 but three dramatically different behaviors near the fixed point.

EXAMPLE 6.6.
a) Let $g(x) = x - x^3$. Then 0 is a fixed point of g and $g'(0) = 1$. Graphical analysis easily demonstrates that $(-1, 1) \subset W^s(0)$.

b) Let $p(x) = x + x^3$. Then 0 is again a fixed point of p and $p'(0) = 1$. However, in this case $W^s(0) = 0$ and $W^s(\infty)$ contains all real numbers except 0.

c) Let $k(x) = e^x - 1$. In this case, 0 is a fixed point of k and $k'(0) = 1$, but $W^s(0) = (-\infty, 0]$ and $W^s(\infty) = (0, \infty)$. That is, all points to the left of 0 approach 0 under iteration of k, but all other points grow without bound under iteration of k.

The graphs of the functions in this example are shown in Figure 6.2. □

Example 6.6 demonstrates that nothing definitive can be said about the behavior of points near a fixed point whose derivative is equal to one in

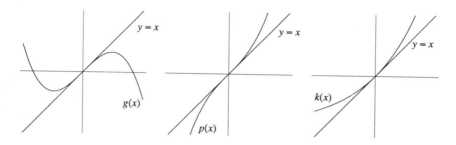

FIGURE 6.2. The graphs of $g(x) = x - x^3$, $p(x) = x + x^3$, and $k(x) = e^x - 1$.

absolute value without further analysis. The reader is invited to investigate such points in Exercise 6.10.

Recall that p is a periodic point with period n of the function f if p is a fixed point of f^n. In light of this fact, the following definition of a hyperbolic periodic point is an obvious extension of the definition of hyperbolic fixed points.

DEFINITION 6.7. *Let f be a differentiable function and p be a periodic point of f with prime period k. If $|(f^k)'(p)| \neq 1$, then p is a hyperbolic periodic point. If $|(f^k)'(p)| = 1$, then p is a nonhyperbolic or neutral periodic point.*

Notice that if p is a periodic point of f with period k, then it is the derivative of f^k at p that is important in determining whether or not p is hyperbolic rather than the derivative of f at p. In Exercise 6.3, we show that if the prime period of p is larger than 1, then the derivative of f at p by itself is not sufficient to determine the behavior of points near p under iteration of the function.

As a corollary to Theorem 6.5, we get

THEOREM 6.8. *Let f be a C^1 function and p be a periodic point of f with prime period k. If $|(f^k)'(p)| < 1$, then there exists an open interval containing p that is contained in the stable set of p. If $|(f^k)'(p)| > 1$, then there exists an open interval containing p such that all points in the interval except p must leave the interval under iteration of f^k.*

Theorem 6.8 is an easy extension of Theorem 6.5; its proof is left as an exercise.

We have observed that whenever p is a periodic point of f with prime period k and the derivative of f^k at p is less than one in absolute value, then points near p are attracted to p under iteration of f^k. On the other hand, if p is a periodic point of f with prime period k and the derivative of f^k at p is greater than one in absolute value, then points near p are moved away from p under iteration of f^k. Hence, we make the following definition.

DEFINITION 6.9. *Let p be a periodic point of f with prime period k. If $|(f^k)'(p)| < 1$, then p is an attracting periodic point of f. If $|(f^k)'(p)| > 1$, then p is a repelling periodic point of f.*

We emphasize again that it is the derivative of f^k at p that is important in determining whether or not p is attracting or repelling. The derivative of f at p by itself provides no information about the nature of the periodic point. (See Exercises 6.2 and 6.3.)

It is true that the fixed point at 0 of the function $g(x) = x - x^3$ attracts nearby points (Example 6.6a). However, $g'(0) = 1$, so we call 0 a *weakly attracting* fixed point of g. Similarly, 0 is a fixed point of $p(x) = x + x^3$, which repels nearby points, but $p'(x) = 1$ (Example 6.6b). Consequently, we call 0 a *weakly repelling* fixed point of p. We investigate methods of determining when a neutral fixed point is weakly attracting or repelling in Exercise 6.10.

In Chapter 7, we will begin to look at how periodic points in particular, and the dynamics of functions in general, change as we slowly change the functions. We shall see that hyperbolic periodic points are relatively insensitive to small changes while nonhyperbolic periodic points can change quickly. The relative stability of hyperbolic periodic points accounts for their importance.

Exercise Set 6

6.1 Find the periodic points of each of the following functions. Indicate whether the periodic points are repelling, attracting, or neither, and find their stable sets.

a) $f(x) = -x^3$

b) $p(x) = x^3 - x$

c) $k(x) = -x^3 - x$

d) $g(x) = e^{x-1}$

e) $E(x) = e^{-x}$

f) $S(x) = \sin x$

g) $r(x) = -x^{1/3}$

h) $A(x) = \frac{-4}{\pi} \arctan x$

•6.2 a) Let $f : I \to I$ be a differentiable function, x be a point in I, and k be a natural number. Prove that

$$(f^k)'(x) = f'(x) \cdot f'(f(x)) \cdot f'(f^2(x)) \cdot \ldots \cdot f'(f^{k-1}(x)).$$

Hint: Use the chain rule and mathematical induction.

b) Let $\{p_1, p_2, \ldots, p_n\}$ be the orbit of a periodic point with period n. Use part (a) to prove p_1 is an attracting hyperbolic periodic point if and only if $|f'(p_1) \cdot f'(p_2) \cdot \ldots \cdot f'(p_n)| < 1$. State and prove a similar statement for repelling hyperbolic periodic points.

c) Let p be a periodic point with prime period k, and let q be in the orbit of p. That is, suppose that $q = f^n(p)$ for some n. Prove that $(f^k)'(p) = (f^k)'(q)$.

d) Show that if p_1 and p_2 are periodic points in the same orbit, then p_1 is attracting if and only if p_2 is.

6.3 Find a C^1 function $f : \mathbb{R} \to \mathbb{R}$ that has 0 as an attracting hyperbolic periodic point with period 2 and such that $f'(0) = 2$. Does it still make sense to call this point attracting? Why or why not?

Hint: Use Exercise 6.2 and draw a graph of f. Revisiting Exercise 4.13 may also be helpful.

6.4 Find a fixed point of $f(x) = -\frac{1}{2}x + 3$ and use the Mean Value Theorem to prove that every real number is in the stable set of this fixed point.

6.5 Find a fixed point of $f(x) = 2x - 1$ and use the Mean Value Theorem to show that every real number except this fixed point is in the stable set of infinity.

6.6 Prove the second half of Theorem 6.5 using the following steps:

a) Use Lemma 2.10 to demonstrate that there exists a $\lambda > 1$ and $\delta > 0$ such that $|f'(x)| > \lambda$ whenever x is in $(p - \delta, p + \delta)$.

b) Use the Mean Value Theorem and mathematical induction to show that if $y \neq p$ and $f^k(y)$ is in $(p - \delta, p + \delta)$ for all $k < n$, then $|f^n(y) - p| > \lambda^n |y - p|$.

c) Use (b) to prove the result.

6.7 Let I be a closed interval and $f : I \to \mathbb{R}$ be a C^1 function satisfying $f(I) \supset I$ and $|f'(x)| < 1$ for all x in I. Theorem 6.3 states that f has a unique fixed point in I. Prove that I is contained in the stable set of this fixed point.

*6.8 Let f be a C^1 function and suppose that p is a fixed point of f satisfying $f'(p) > 1$. Does Theorem 6.5 imply that there is a neighborhood U of p such that $U \cap W^s(p) = p$? Prove that your answer is correct.

6.9 Let $f : I \to \mathbb{R}$ be a C^1 function of the interval I, and suppose that p is a fixed point of f.

a) Prove that if $f'(p) < 0$, then there is an open interval U containing p such that whenever x is in U, $f(x)$ and x are on opposite sides of p.

b) Prove that if $f'(p) > 0$, then there is an open interval U containing p such that whenever x is in U, $f(x)$ and x are on the same side of p.

Hint: First, define mathematically what we mean by "being on the same side of p", and then use the Mean Value Theorem. You may also wish to use Lemma 2.10.

6.10 AN INVESTIGATION—NEUTRAL FIXED POINTS:

In Example 6.6 we considered three functions that had fixed points at which the derivative was equal to 1. In the first case the fixed point was weakly attracting, in the second it was weakly repelling, and in the third it was neither. Examine the graphs of these functions in Figure 6.2. Can you describe what the graph of the function should look like for each type of neutral fixed point? Can you tell what the second derivative should be in each case? Check your conjectures using functions of the form $p(x) = x \pm x^n$.

Let $f(x)$ be a differentiable function with derivatives of all orders. Suppose p is a fixed point and $f'(p) = 1$. State and prove a theorem that would determine whether or not p is weakly attracting, weakly repelling, or neither based on the sign and order of the first nonzero derivative of f at p with order higher than one.

* Let $f(x)$ be a differentiable function with derivatives of all orders. Suppose p is a fixed point and $f'(p) = -1$. State and prove a theorem that would determine whether or not p is weakly attracting, weakly repelling, or neither based on the sign and order of the first nonzero derivative of f at p with order higher than one.

6.11 Let x_0 and α be positive real numbers and consider the sequence defined by
$$x_{n+1} = \frac{1}{2}\left(x_n + \frac{\alpha}{x_n}\right).$$

a) Find a function f such that $x_n = f^n(x_0)$.

b) Describe the periodic points of f. Are they hyperbolic? Are they attracting or repelling?

c) What is the limit of the sequence $x_0, x_1, x_2, \ldots$?

Note: The function f found in part (a) provides a very useful algorithm for computing square roots since it has a strongly attracting fixed point. That is, the sequence converges quickly to the desired square root. For a discussion of the error estimate, see Exercise 16 in Chapter 3 of the third edition of W. Rudin's text, *Principles of Mathematical Analysis*.

7
Parametrized Families of Functions and Bifurcations

By a family of functions we mean a collection of functions that are of the same general type. For example we consider the family of linear functions of the form $f_m(x) = mx$, where m is allowed to vary over the real numbers. The variable m is called a *parameter* and the family represented by $f_m(x) = mx$ is called a *parametrized family*. The parameter is usually written as a subscript of the function name. In those cases where the parameter is clear, we will suppress the subscript to simplify notation.

We are interested in how the dynamics of a function changes as we change the parameter. To illustrate this, let's look at an example.

EXAMPLE 7.1.
Consider the parametrized family $f_m(x) = mx$. For all values of m except 1, it is clear that 0 is the only fixed point of f_m.

If $m < -1$, then 0 is a hyperbolic repelling fixed point and all other points are in the stable set of infinity.

If $m = -1$, then all points except 0 are periodic points with prime period 2.

When $-1 < m < 1$, 0 is an attracting fixed point and $W^s(0) = \mathbb{R}$.

If $m = 1$, then all points are fixed points.

Finally, when $m > 1$, 0 is a repelling fixed point and all other points are in the stable set of ∞. □

We note that the dynamics of the family $f_m(x) = mx$ are unchanged for

60   7. Parametrized Families of Functions and Bifurcations

large intervals of parameter values. Then, at a particular parameter value (i.e., -1 and 1), the dynamics change suddenly, after which they again remain constant for a prolonged interval. We call these sudden changes in dynamics *bifurcations*. Notice also that 0 is the fixed point and that $|f'(0)| = 1$ when $m = 1$ or -1. The presence of nonhyperbolic fixed points for parameter values at which there is a bifurcation is typical.

DEFINITION 7.2. *Let $f_c(x)$ be a parametrized family of functions. Then there is a bifurcation at c_0 if there exists $\epsilon > 0$ such that whenever a and b satisfy $c_0 - \epsilon < a < c_0$ and $c_0 < b < c_0 + \epsilon$, then the dynamics of $f_a(x)$ are different from the dynamics of $f_b(x)$. In other words, the dynamics of the function changes when the parameter value crosses through the point c_0.*

Readers who are used to a more rigorous approach may wonder what we mean by saying "the dynamics are the same" or the "dynamics are different". We will rely on an intuitive idea of sameness until we discuss topological conjugacy in Chapter 9. Then we will say two functions have the same dynamics if and only if they are topologically conjugate. Interested readers are encouraged to look ahead. For now, our intuition will be able to deal well with the functions we examine.

Let's look at a few more simple examples.

EXAMPLE 7.3.
Consider the family of functions $E_c(x) = e^{x+c}$. We examine the dynamics of this family by using graphical analysis. We are all familiar with the general shape of the exponential function. We should notice that as the parameter increases in value the shape of the graph remains the same, but it moves to the left. A careful examination reveals that we have three cases: the graphs of $E(x)$ and $y = x$ intersect in exactly two points, the graphs of $E(x)$ and $y = x$ have a single point of tangency, and the graphs of $E(x)$ and $y = x$ do not intersect. The graphs of typical examples are shown in Figure 7.1.

If $c < -1$, then the graphs of $E(x)$ and $y = x$ intersect in two places, as shown in Figure 7.1. In the figure, we label the two points of intersection a and b, with $a < b$. Note that both a and b are fixed points, $0 < E'(a) < 1$, and $E'(b) > 1$. Hence, a is an attracting hyperbolic fixed point and b is a repelling hyperbolic fixed point. Graphical analysis demonstrates that $W^s(a) = (-\infty, b)$ and $W^s(\infty) = (b, \infty)$.

When $c = -1$, the two fixed points are fused into one fixed point at $x = 1$. This fixed point is *not* hyperbolic since $E'_{-1}(1) = 1$. Graphical analysis demonstrates that $W^s(1) = (-\infty, 1]$ and $W^s(\infty) = (1, \infty)$. This case is also illustrated in Figure 7.1.

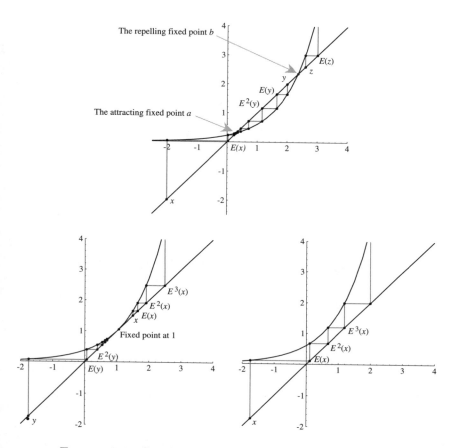

FIGURE 7.1. Graphical analysis of E_c for the three cases $c < -1$, $c = -1$, and $c > -1$.

Finally, if $c > -1$, then the graphs of $E(x)$ and $y = x$ do not intersect, and consequently $E(x)$ has no fixed point. Since $E(x)$ is continuous, Sarkovskii's Theorem implies that $E(x)$ can have no periodic points when $c > -1$. Using graphical analysis, we show in Figure 7.1 that $W^s(\infty) = \mathbb{R}$. □

We make a special note of a few characteristics of this example. First, as the parameter c grows and approaches -1, the two fixed points a and b gradually approach one another until, when $c = -1$, they join to become one fixed point. Immediately thereafter, they disappear altogether. This type of bifurcation is called a *saddle-node bifurcation*. The merging and annihilation (or creation and splitting) of periodic points as we vary a

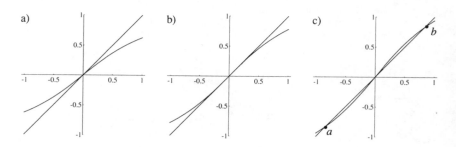

FIGURE 7.2. The graph of $A_k(x) = k \arctan(x)$ along with the line $y = x$ for the parameters values a) $k = .8$, b) $k = 1$, and c) $k = 1.2$. Note that $A'_k(0) = k$ when $k = 1$, so the fixed point is not hyperbolic.

parameter is a common theme in dynamics. Further, notice that at the moment of bifurcation, the fixed point is not hyperbolic. This is another common characteristic of bifurcations.

EXAMPLE 7.4.

Consider the family of functions $A_k(x) = k \arctan(x)$ for parameter values near 1. When $0 < k < 1$, the function has one attracting fixed point near 0 and all real numbers are in the stable set of 0. If $k = 1$, then 0 is still a fixed point but $A'_1(0) = 1$. Graphical analysis indicates that 0 is weakly attracting and all real numbers are in the stable set of 0 even though 0 is not hyperbolic. Finally, when $k > 1$, we find a repelling fixed point at 0 ($A'_k(0) = k > 1$) and two other fixed points have been formed, both of which are attracting. If the two fixed points are labeled a and b and $a < 0 < b$, then we see that $W^s(a) = (-\infty, 0)$ and $W^s(b) = (0, \infty)$. The graph of each case is shown in Figure 7.2 along with the line $y = x$. □

The type of bifurcation seen in $A_k(x) = k \arctan(x)$ when $k = 1$ is called a *pitchfork bifurcation*. In general, pitchfork bifurcations occur when either an attracting periodic point splits into a repelling periodic point with an attracting periodic point of the same period as the original point on each side of it, or alternatively, a repelling periodic point splits into an attracting periodic point surrounded by two repelling periodic points of the same period. The rationale for this odd name becomes apparent when we consider the bifurcation diagram.

A *bifurcation diagram* is a graph of the periodic points of a function plotted as a function of the parameter. Typically, the periodic values are plotted on the vertical axis and the parameter values are plotted on the

7. Parametrized Families of Functions and Bifurcations

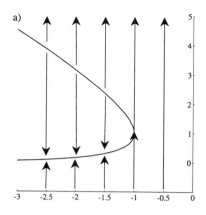
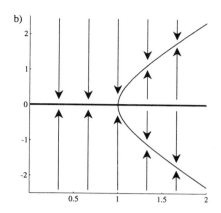

FIGURE 7.3. The bifurcation diagrams of a) $E_c(x) = e^{x+c}$ and b) $A_k(x) = k\arctan(x)$. Note that the value of the parameter is shown on the x-axis and the domain of the function is represented on the y-axis. Solid lines represent fixed points. From graph (a) we see that when $c = -2$, $E_c(x)$ has fixed points at about .16 and about 3.2. The arrows indicate the direction points move under iteration. Consequently, when $c = -2$ the arrows indicate the fixed point at .16 is attracting and the fixed point at 3.2 is repelling. Recall that bifurcations whose diagrams are similar to (a) are called saddle node bifurcations and those whose diagrams are similar to (b) are called pitchfork bifurcations. The inspiration for the latter name should be apparent from the diagram.

horizontal axis. We denote fixed points by solid lines and periodic points that are not fixed by dotted lines. Vertical arrows are added to indicate whether or not a point is attracting or repelling. Bifurcation diagrams for $E_c(x) = e^{x+c}$ and $A_k(x) = k\arctan(x)$ are shown in Figure 7.3. A quick look at the bifurcation diagram for A_k attests to the origin of the name "pitchfork bifurcation".

EXAMPLE 7.5.
  The logistic function $h_r(x) = rx(1-x)$ exhibits a third type of bifurcation when $r = 1$: a *transcritical bifurcation*. When $0 < r < 1$, h has two fixed points, one of which is less than one and repelling, and another at 0 that is attracting. (To locate the fixed points, we solve the equation $h_r(x) = x$. We find that there are fixed points at 0 and $p_r = \frac{r-1}{r}$.)

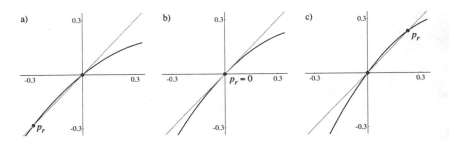

FIGURE 7.4. The graph of $h_r(x) = rx(1-x)$ for three parameter values near 1. The graph of the line $y = x$ is also shown for comparison. Notice the fixed points at 0 and p_r. In graph (a) $r = .78$, in (b) $r = 1$, and in (c) $r = 1.3$. Note also that $h_1'(0) = 1$ when $r = 1$, so the fixed point is not hyperbolic when the bifurcation occurs.

When $r = 1$, these two points merge to form one fixed point at 0. This non-hyperbolic fixed point attracts numbers greater than 0 and repels points less than 0. Finally, if $1 < r < 3$, we see that 0 becomes a repelling fixed point and there is an attracting fixed point that is greater than 0. The graphs of these three situations are shown in Figure 7.4. □

EXAMPLE 7.6.
Continuing the investigation of $h_r(x) = rx(1-x)$ that we started in Example 7.5, we find that h_r undergoes another bifurcation when $r = 3$. We have seen that h_r has a repelling fixed point at 0 and an attracting fixed point greater than 0 when $1 < r < 3$. A quick look at the graph of h_r^2 demonstrates that h_r has no other periodic points when r is in this range. At $r = 3$, the larger fixed point is at .75 and is weakly attracting; it is no longer hyperbolic because $h'(.75) = -1$. When $r > 3$, the fixed point is repelling. (The proof of this is left as an exercise.) Graphical analysis when $r = 3.1$ indicates that most points in $(0, 1)$ are attracted to a period two orbit that straddles the larger fixed point. What has happened?

To understand this, we look at the graph of h_r^2 in the vicinity of the fixed point $p_r = \frac{r-1}{r}$ for parameter values slightly less than 3, equal to 3, and slightly larger than 3. Some examples are shown in Figure 7.5. Notice that as the parameter value passes 3, the value of $(h_r^2)'(p_r)$ changes from being less than 1 to being greater than 1. Hence, the fixed point changes from attracting to repelling. In addition, the continuity of h_r requires that as this happens a period two attracting orbit must be added. (The details

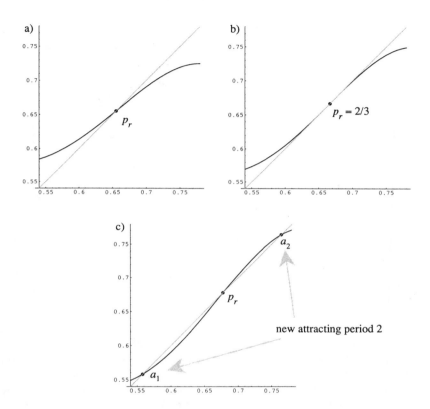

FIGURE 7.5. The graph of $h_r^2(x)$ where $h_r(x) = rx(1-x)$ for three parameter values. In (a) $r = 2.9$, in (b) $r = 3$, and in (c) $r = 3.1$. Note that the attracting fixed point splits into a period two attracting orbit with a repelling fixed point between the two points in the orbit. The period two orbit is indicated by a dashed line in the bifurcation diagram shown in Figure 7.6.

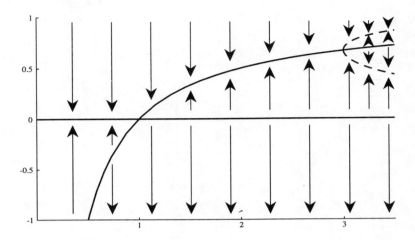

FIGURE 7.6. The bifurcation diagram for the logistic function $h_r(x) = rx(1-x)$. The dashed lined represents a period 2 attracting orbit.

of the proof of this fact are also left as an exercise.) This orbit is labeled by the points a_1 and a_2 in Figure 7.5. This type of bifurcation is called a *period-doubling bifurcation* since a periodic orbit of twice the period of the original periodic point is added. A bifurcation diagram for the family h_r is shown in Figure 7.6. □

In closing, we note that there are several ways of drawing bifurcation diagrams. The choice of diagram depends on the information we are trying to code. We will see another type of bifurcation diagram in Chapter 10. We also note that of the four types of bifurcations described here, the most important are the saddle-node and the period-doubling bifurcations. They arise in many settings and can't be eradicated by small changes in the parametrized family. Transcritical and pitchfork bifurcations are interesting but are only observed in certain types of parametrized families, and they can be eradicated by small changes in the family.

Exercise Set 7

7.1 Discuss the bifurcations of the following families of functions at the given parameter values and draw bifurcation diagrams for each.

Hint: It may help to begin by drawing the graph of the function along with the line $y = x$ for parameter values close to the bifurcation value.

a) $q_c(x) = x^2 + c$ at $c = \frac{1}{4}$ and $-\frac{3}{4}$.

b) $F_r(x) = rx - x^3$ at $r = 1$.

c) $E_c(x) = e^x + c$ at $c = -1$.

d) $g_m(x) = x^3 + mx$ at $m = 1$ and -1.

e) $E_k(x) = e^{k(x-1)}$ at $k = 1$.

7.2 Let $h_r(x) = rx(1-x)$.

a) Show that the only fixed points of h_r are 0 and $\frac{r-1}{r}$.

b) Use the derivative of h_r to determine for which parameter values the fixed points of h_r are repelling and for which values they are attracting.

7.3 In Example 7.6, we stated that a period two orbit must be added as the derivative of h_r^2 at $p_r = \frac{r-1}{r}$ changed from being less than one to being greater than one. Prove that this is true.

Hint: To develop an intuitive understanding of the problem, begin by plotting a few graphs of $h_r(x)$ on $[0, 1]$ for a few parameter values near the bifurcation. Then show that $(h_r^2(p_r))' > 1$ implies that there exists $x_0 > p_r$ satisfying $h_r^2(x_0) > x_0$. Use the fact that $h_r^2(1) = 0$ and the Intermediate Value Theorem to show that there must exist a fixed point of h_r^2 in the interval $(p_r, 1)$. Using a similar technique, show that there exists a fixed point of h_r^2 in $(\frac{1}{2}, p_r)$. How do we know that these points are prime period two points of h_r? Can you prove that they are in the same orbit?

7.4 AN INVESTIGATION: Examine the dynamics of h_r for the parameter values between 0 and 4. Find as many of the bifurcations as you can and describe each one.

8

The Logistic Function
Part I: Cantor Sets and Chaos

We return to the family of functions $h_r(x) = rx(1-x)$ where $r > 0$. As we saw in Chapter 1, this family is a reasonable model of population growth. We note that the graph of h_r is a parabola facing down with x intercepts at 0 and 1, and with a vertex at the point $(\frac{1}{2}, \frac{r}{4})$.

The following facts have already been discussed or are easily verified:

(1) Solving the equation $rx(1-x) = x$ demonstrates that h_r has fixed points at 0 and at $p_r = \frac{r-1}{r}$. Also, 1 and $\frac{1}{r}$ are eventually fixed points since $h_r(1) = 0$ and $h_r(\frac{1}{r}) = p_r$.

(2) If $0 < r < 1$, then $p_r < 0$, p_r is a repelling fixed point and 0 is an attracting fixed point. The stable set of 0 is $(p_r, \frac{1}{r})$, the stable set of p_r is $\{p_r, \frac{1}{r}\}$, and the stable set of infinity is $(-\infty, p_r) \cup (\frac{1}{r}, \infty)$.

(3) When $r = 1$, a transcritical bifurcation occurs, $p_1 = 0$ and is not hyperbolic, the stable set of 0 is $[0, 1]$, and the stable set of infinity is $(-\infty, 0) \cup (1, \infty)$. In general, if $r \geq 1$, then the stable set of infinity includes the intervals $(-\infty, 0)$ and $(1, \infty)$.

(4) If $1 < r < 3$, then 0 is a repelling fixed point, p_r is an attracting fixed point, the stable set of 0 is $\{0, 1\}$, and the stable set of p_r is $(0, 1)$.

(5) When $r = 3$, a period-doubling bifurcation occurs and p_r is only weakly attracting, but the fixed point at 0 is still repelling. The stable set of p_r is still $(0, 1)$.

(6) If $3 < r < 3.4$, then both 0 and p_r are repelling fixed points. Note that the stable set of 0 is still $\{0, 1\}$, but the stable set of p_r contains an infinite number of points. However, there is now an attracting period two orbit and "most" of the points in $(0, 1)$ are forward asymptotic to one of the two points in the orbit.

(7) When $r \approx 3.45$, another period-doubling bifurcation occurs and the period two attracting orbit splits into a period four attracting orbit and a period two repelling orbit.

(8) In the interval $3.44 < r \leq 4$ changes occur rapidly. There is a period three orbit whenever $r > 3.6$, so by Sarkovskii's Theorem we know that there are periodic points of all orders. Because of the complexity of the dynamics when r is in this range, we will first discuss the dynamics when $r > 4$. The behavior when $r = 4$ will be analyzed in Chapter 9, and the dynamics in the range $3.44 < r < 4$ is examined in Chapter 10.

8.1. A First Look at the Logistic Function when $r > 4$

Suppose that $r > 4$ and $h_r(x) = rx(1-x)$. Then $h(\frac{1}{2}) > 1$ and $\frac{1}{2}$ is forward asymptotic to infinity. In addition, since $h(\frac{1}{2}) > 1$, $h(1) = 0$, and $h(0) = 0$, the Intermediate Value Theorem implies that there exist x_0 in $[0, \frac{1}{2}]$ and x_1 in $[\frac{1}{2}, 1]$ such that $h(x_0) = h(x_1) = 1$. Consequently, $h^2(x_0) = h^2(x_1) = 0$ and both x_0 and x_1 are eventually fixed at 0 by h. (See Figure 8.1 for an illustration of this arrangement.) We leave as an exercise the proof that there are infinitely many points that are eventually fixed at 0 and that there are infinitely many points that are eventually fixed at p_r. We would like to know what other points of $[0, 1]$ are forward asymptotic to something in $[0, 1]$.

For each natural number n, we define

$$\Lambda_n = \{x \mid h^n(x) \text{ is in } [0, 1]\}. \tag{8.1}$$

Our goal is to describe the set $\Lambda = \bigcap_{n=1}^{\infty} \Lambda_n$: the set of points that remain in $[0, 1]$ forever under iteration of h. To do this, we need to know more about the sets Λ_n. Throughout the remainder of this book, the symbol Λ_n will be used to indicate the set of points that remain in $[0, 1]$ after n iterations of h. The symbol Λ will be consistently used to indicate the intersection of the sets Λ_n. The following proposition is a first step in developing a description of Λ.

8.1. A First Look at the Logistic Function when $r > 4$

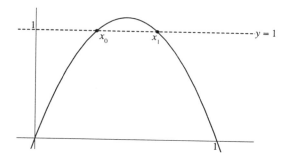

FIGURE 8.1. The graph of $h(x) = rx(1 - x)$ when $r > 4$. The points where $h(x) = 1$ are $x_0 = \frac{1}{2} - \frac{\sqrt{r^2 - 4r}}{2r}$ and $x_1 = \frac{1}{2} + \frac{\sqrt{r^2 - 4r}}{2r}$. The vertex is at $(\frac{1}{2}, \frac{r}{4})$.

PROPOSITION 8.1. *If $h(x) = rx(1 - x)$ and $r > 4$, then the following statements are true.*

a) *The set of real numbers x in $[0, 1]$ satisfying the condition that $h(x)$ is not in $[0, 1]$ is the interval*

$$\left(\frac{1}{2} - \frac{\sqrt{r^2 - 4r}}{2r}, \frac{1}{2} + \frac{\sqrt{r^2 - 4r}}{2r}\right).$$

Further,

$$\Lambda_1 = \left[0, \frac{1}{2} - \frac{\sqrt{r^2 - 4r}}{2r}\right] \cup \left[\frac{1}{2} + \frac{\sqrt{r^2 - 4r}}{2r}, 1\right].$$

b) *The set Λ_n consists of 2^n disjoint closed intervals for all natural numbers n.*

c) *If I is one of the 2^n closed intervals in Λ_n, then $h^n : I \to [0, 1]$ is one-to-one and onto.*

PROOF. We first note that if $h(x)$ is in $[0, 1]$, then x must be in $[0, 1]$. Hence, Λ_n is contained in $[0, 1]$ for all n. We now prove part (a) of the proposition.

The graph of $h(x)$ is shown in Figure 8.1. We see that the set of points in $[0, 1]$ for which $h(x)$ is not in $[0, 1]$ contains exactly those points that satisfy $h(x) > 1$. But these are the points that lie between the roots of $h(x) = 1$. Using the quadratic formula, we find that $h(x) = rx(1 - x) = 1$ when $x = \frac{1}{2} \pm \frac{\sqrt{r^2 - 4r}}{2r}$ and (a) follows.

We prove (b) and (c) by induction. It is clear from part (a) that Λ_1 consists of 2^1 disjoint closed intervals. Also, we see that $h(0) = h(1) = 0$

and $h(\frac{1}{2} \pm \frac{\sqrt{r^2-4r}}{2r}) = 1$. Thus, the endpoints of the two intervals in Λ_1 are mapped to 0 and 1. By continuity and the Intermediate Value Theorem, we can conclude that if I is one of the two intervals in Λ_1, then $h : I \to [0,1]$ is onto. To verify that h is one-to-one on the intervals of Λ_1, we note that it is strictly monotone on these intervals since $h'(x) = r(1-2x) > 0$ when $x < \frac{1}{2}$ and $h'(x) < 0$ when $x > \frac{1}{2}$. Therefore, if I is one of the two intervals in Λ_1, then $h : I \to [0,1]$ is one-to-one and onto.

To continue the induction argument, we suppose that Λ_k consist of 2^k disjoint closed intervals. Further, we suppose that if $[a,b]$ is one of the closed intervals comprising Λ_k, then $h^k : [a,b] \to [0,1]$ is one-to-one and onto and that either $(h^k)'(x) < 0$ for all x in $[a,b]$ or $(h^k)'(x) > 0$ for all x in $[a,b]$.

Now consider Λ_{k+1}. It is easily seen that $\Lambda_{k+1} \subset \Lambda_k$; the reader should be able to explain why this is so. Our task then is to ascertain what portions of Λ_k are also in Λ_{k+1}. To do this, we let $[a,b]$ be one of the 2^k intervals that comprise Λ_k. Now either $(h^k)'(x) > 0$ for all x in $[a,b]$ or $(h^k)'(x) < 0$ for all x in $[a,b]$. In the first case, h^k is strictly increasing on $[a,b]$ and in the second h^k is strictly decreasing on $[a,b]$. We begin by assuming that $(h^k)'(x) > 0$ for all x in $[a,b]$. Since h^k is strictly increasing on $[a,b]$, h^k is continuous, and $h^k([a,b]) = [0,1]$, the Intermediate Value Theorem implies that there exist unique points x_2 and x_3 satisfying

(1) $a < x_2 < x_3 < b$,
(2) $h^k([a,x_2]) = [0, \frac{1}{2} - \frac{\sqrt{r^2-4r}}{2r}]$,
(3) $h^k((x_2,x_3)) = (\frac{1}{2} - \frac{\sqrt{r^2-4r}}{2r}, \frac{1}{2} + \frac{\sqrt{r^2-4r}}{2r})$, and
(4) $h^k([x_3,b]) = [\frac{1}{2} + \frac{\sqrt{r^2-4r}}{2r}, 1]$.

The first condition implies that the intervals $[a, x_2]$ and $[x_3, b]$ are disjoint. The last three conditions imply that $h^{k+1}([a,x_2]) = [0,1]$, $h^{k+1}(x) > 1$ for all x in (x_2, x_3), and $h^{k+1}([x_3, b]) = [0,1]$. Thus, the set of points in $[a,b]$ that are also in Λ_{k+1} consists of the two disjoint closed intervals $[a,x_2]$ and $[x_3,b]$. If x is in $[a,x_2]$, then $h^k(x)$ is in $[0, \frac{1}{2} - \frac{\sqrt{r^2-4r}}{2r}]$ and $h'(h^k(x)) > 0$. By assumption, $(h^k)'(x) > 0$, so we are assured that

$$(h^{k+1})'(x) = h'(h^k(x)) \cdot (h^k)'(x) > 0$$

for all x is in $[a, x_2]$. A similar argument demonstrates $(h^{k+1})'(x) < 0$ for all x in $[x_3, b]$.

If we begin the argument in the preceding paragraph by assuming that $(h^k)'(x) < 0$ for all x in $[a,b]$, then we can use an analogous argument to show that the points of Λ_{k+1} that are in $[a,b]$ are contained in two disjoint intervals $[a, x_2]$ and $[x_3, b]$, $h^{k+1}([a, x_2]) = [0,1]$, $h^{k+1}([x_3, b]) = [0,1]$ and that $(h^{k+1})'$ is either strictly positive or strictly negative on each of the intervals $[a, x_2]$ and $[x_3, b]$. Since $[a,b]$ is an arbitrary interval in Λ_k, it follows that there are twice as many intervals in Λ_{k+1} than in Λ_k. That is, Λ_{k+1} contains $2(2^k) = 2^{k+1}$ disjoint closed intervals. Further, we have shown that if J is one of the intervals in Λ_{k+1}, then $h^{k+1} : J \to [0,1]$ is onto (by conditions (2) and (4) listed earlier). Also, h^{k+1} is one-to-one on J, and we can make the next step in the induction since $(h^{k+1})'$ is either strictly positive or strictly negative on J. □

As we stated earlier, our goal is to describe the set $\Lambda = \bigcap_{n=1}^{\infty} \Lambda_n$. In the proof of Proposition 8.1, we saw that Λ_{k+1} was formed by removing an open interval from the middle of each interval in Λ_k. The set of points left after this has been done for each n is Λ. Students of topology may recognize that this construction is similar to that of a Cantor middle-thirds set.

8.2. Cantor Sets

DEFINITION 8.2. *A nonempty set $\Gamma \subset \mathbb{R}$ is called a Cantor set if*

a) *Γ is closed and bounded. (Sets of real numbers with these characteristics are called compact sets.)*

b) *Γ contains no intervals. (Sets of this nature are called totally disconnected sets.)*

c) *Every point in Γ is an accumulation point of Γ. (When closed, such sets are called perfect sets.)*

While most of the terms used in Definition 8.2 should be familiar, we have not yet defined bounded sets. A set of real numbers is *bounded* if there exists a positive number that is larger than the absolute value of every element in the set. This is equivalent to saying that there exists a closed interval that contains the set.

EXAMPLE 8.3. CONSTRUCTING MIDDLE-α CANTOR SETS.

Cantor sets are usually constructed by iterative processes. To construct the Cantor middle-thirds set, we begin with the interval $[0, 1]$ and remove the open set $(\frac{1}{3}, \frac{2}{3})$. In the second step, we remove the middle third of each remaining interval. That is, we remove the intervals $(\frac{1}{9}, \frac{2}{9})$ and $(\frac{7}{9}, \frac{8}{9})$. We continue the process by removing the middle third of each of the remaining intervals at each step. The set of points remaining is called the *Cantor middle-thirds set*. As we will see shortly, this is in fact a Cantor set.

If $0 < \alpha < 1$, then we can construct similar sets called *Cantor middle-α sets* by removing an interval from the center of each remaining interval whose length is α times the length of the remaining interval. For example, in the middle-$\frac{1}{5}$ set we remove the middle fifth of each interval at each step.

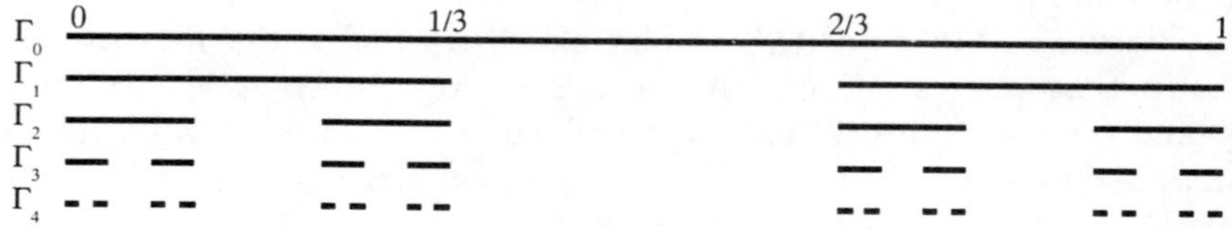

```
Γ₀  0                    1/3                  2/3                  1
    ─────────────────────────────────────────────────────────────────
Γ₁  ──────────────────────────              ──────────────────────────
Γ₂  ─────────      ─────────                ─────────       ─────────
Γ₃   ───   ───      ───   ───                ───   ───       ───   ───
Γ₄   ‑‑ ‑‑ ‑‑ ‑‑    ‑‑ ‑‑ ‑‑ ‑‑              ‑‑ ‑‑ ‑‑ ‑‑     ‑‑ ‑‑ ‑‑ ‑‑
```

FIGURE 8.2. The first five sets formed in the construction of the Cantor middle-thirds set.

To be more precise, let $\Gamma_0 = [0,1]$. Define Γ_1 to be the two closed intervals of equal length left when the open interval of length α is removed from the middle of Γ_0. Define Γ_2 to be the set of four closed intervals of equal length obtained by removing an open interval whose length is $\alpha \times$ (the length of an interval in Γ_1) from the middle of each closed interval in Γ_1. We continue to define Γ_n inductively as the set of closed intervals of equal length formed by removing an open interval whose length is $\alpha \times$ (the length of an interval in Γ_{n-1}) from the center of the closed intervals in Γ_{n-1}. The first few iterations of this construction for $\alpha = \frac{1}{3}$ are shown in Figure 8.2. The Cantor middle-α set is the set $\Gamma = \bigcap_{n=0}^{\infty} \Gamma_n$.

Clearly, Γ is not empty since the endpoints of the open intervals removed are in Γ. We shall see in Proposition 8.5 that every point in Γ is an accumulation point of Γ. Thus Γ is perfect since Proposition 3.15 implies that Γ is closed. Students who have studied countability will be interested to know that nonempty perfect subsets in $\mathbb{R}$ must be uncountable. This is proven in the analysis text by Rudin, which is listed in the references. It is also interesting to note that only countably many of the points in Γ are endpoints of the intervals that are removed, so there are uncountably many points in Γ that are not endpoints. □

LEMMA 8.4. *If Γ_n is as defined in Example 8.3, then there are 2^n closed intervals in Γ_n and the length of each closed interval is $\left(\frac{1-\alpha}{2}\right)^n$. Also, the combined length of the intervals in Γ_n is $(1-\alpha)^n$, which approaches 0 as n approaches infinity.*

PROOF. We start with an interval of length 1 and proceed by mathematical induction.

In the first step, we remove a gap of length α and are left with $2 = 2^1$ closed intervals with a combined length of $(1-\alpha)$. So each interval has length $\frac{1-\alpha}{2}$.

In general, suppose that there are 2^k intervals left in Γ_k, each with a length of $\left(\frac{1-\alpha}{2}\right)^k$ for a combined length of $(1-\alpha)^k$. We will show that there are 2^{k+1} intervals left in Γ_{k+1}, each with length $\left(\frac{1-\alpha}{2}\right)^{k+1}$ for a combined length of $(1-\alpha)^{k+1}$. Note that each time we remove the middle-α portion of a closed interval, we split the interval into two closed intervals. So in passing from Γ_k to Γ_{k+1} we double the number of intervals, and there are $2(2^k) = 2^{k+1}$ intervals in Γ_{k+1}. By assumption, each interval in Γ_k has a length of $\left(\frac{1-\alpha}{2}\right)^k$. Since we removed the middle-α portion of each interval in Γ_k to create Γ_{k+1}, the amount of each interval from Γ_k left in Γ_{k+1} is $\left(\frac{1-\alpha}{2}\right)^k - \alpha\left(\frac{1-\alpha}{2}\right)^k = \left(\frac{(1-\alpha)^{k+1}}{2^k}\right)$. As this length is left in two intervals, the length of each remaining interval is $\frac{1}{2}\left(\frac{(1-\alpha)^{k+1}}{2^k}\right) = \left(\frac{1-\alpha}{2}\right)^{k+1}$. Finally, there are 2^{k+1} intervals in Γ_{k+1}, so the combined length of the intervals in Γ_{k+1} is $(2^{k+1})\left(\frac{1-\alpha}{2}\right)^{k+1} = (1-\alpha)^{k+1}$.

Since $0 < \alpha < 1$, $(1-\alpha)^n$ converges to 0 as n grows without bound, and it follows that the combined length of the intervals in Γ_n approaches 0 as n goes to infinity. □

We are now ready to show that Cantor middle-α sets are appropriately named.

PROPOSITION 8.5. *The Cantor middle-α set is a Cantor set.*

PROOF. Let Γ be a Cantor middle-α set. Since 0 is in every Γ_n, Γ is not empty. To complete the proof, we must show that a) Γ is closed and bounded, b) Γ contains no intervals, and c) every point of Γ is an accumulation point of Γ.

a) Since Γ is the intersection of a set of closed intervals, Proposition 3.15 implies that it is closed. As Γ is contained in $[0,1]$, it is also bounded.

b) If Γ contains an open interval (x,y) with length $|y-x|$, then at each stage in the construction of Γ, (x,y) must be contained in one of the remaining closed intervals. However, Lemma 8.4 implies that after n steps the length of one of these intervals is $\left(\frac{1-\alpha}{2}\right)^n$ and we can find an n_0 such that $\left(\frac{1-\alpha}{2}\right)^{n_0} < |y-x|$. That is, the length of each of the closed intervals in Γ_{n_0} is less than the length of (x,y). Hence, the entire interval (x,y) cannot be contained in Γ_{n_0} and Γ contains no intervals.

c) Suppose that x is a point in Γ and let $N_\epsilon(x) = (x-\epsilon, x+\epsilon)$ be a neighborhood of x. We must show that there exists a point in Γ that is contained in $N_\epsilon(x)$ and is not equal to x. Notice that if a is an endpoint of one of the intervals that is removed, then a is in Γ. Now at each stage in the construction of the Cantor set, x must be in one of the remaining closed intervals. That is, for each n there is an interval in Γ_n that contains x. Choose n large enough so that $\left(\frac{1-\alpha}{2}\right)^n < \epsilon$. Then x is in one of the closed intervals that comprise Γ_n. Call this interval I_n. By Lemma 8.4, the length of I_n is $\left(\frac{1-\alpha}{2}\right)^n$. Since $\left(\frac{1-\alpha}{2}\right)^n < \epsilon$, it must be that the endpoints of I_n are in $N_\epsilon(x)$. As there are two endpoints and x can be equal to at most one of them, we are done. □

Note that the proof of Proposition 8.5 does not use the fact that we removed the α-interval from the exact middle of the closed interval. The fact that the total length of all the intervals approaches zero follows from the observation that if we remove a fraction α of the combined length at each stage, then the length is reduced by a factor of $1 - \alpha$ and the total length of the remaining intervals is $(1 - \alpha)^n$ after n steps. In fact, if at each step we remove an open interval whose length is α times the length of each closed interval and we remove this interval in such a way that the endpoints of the open interval do not coincide with either of the endpoints of the closed interval, then we obtain a Cantor set. It can also be shown that if a subset of the real line is closed and bounded, contains no open intervals, and is perfect, then it is homeomorphic to the Cantor middle-thirds set. A proof of this fact can be found in the topology text by Willard, which is listed in the references.

8.3. Chaos and the Dynamics of the Logistic Function

We continue our exploration of the dynamics of the logistic function by considering the case when $r > 2 + \sqrt{5}$. We show that the set of points that remain in $[0, 1]$ under iteration of h_r is a Cantor set whenever $r > 2 + \sqrt{5}$ and the dynamics of h are chaotic on this Cantor set. This is true for all $r > 4$, but it is a bit trickier to prove for $r < 4 \leq 2 + \sqrt{5}$. Recall from Section 8.1 that the set of points remaining in $[0, 1]$ after n iterations of h is denoted Λ_n and that the set of points that never leave $[0, 1]$ under iteration of h is denoted Λ; that is, $\Lambda = \bigcap \Lambda_n$. To construct the argument that Λ is a Cantor set when $r > 2 + \sqrt{5}$, we use the following lemma.

LEMMA 8.6. *Let $r > 2 + \sqrt{5}$ and $h(x) = rx(1-x)$. Then there is $\lambda > 1$ such that $|h'(x)| > \lambda$ whenever x is in Λ_1. Further, the length of each interval in Λ_n is less than $(\frac{1}{\lambda})^n$.*

Note that the value of λ found in the lemma depends on the value of r. As r gets smaller, then λ must also get smaller.

The proof of the first part of the lemma follows from the fact that if x is in Λ_1, then $|h'_r(x)|$ is greater than or equal to the absolute value of the derivative of h_r at $(\frac{1}{2} \pm \frac{\sqrt{r^2-4r}}{2r})$. We prove the second portion of the lemma by using the first part and the Mean Value Theorem. A proof is outlined in Exercise 8.3.

8.3. Chaos and the Dynamics of the Logistic Function 77

THEOREM 8.7. *If $r > 2 + \sqrt{5}$, then the set $\Lambda = \bigcap_{n=1}^{\infty} \Lambda_n$ is a Cantor set.*

PROOF. Since 0 is a fixed point of h, it is clearly in Λ. Thus, Λ is not empty. To prove that Λ is a Cantor set, we need to show that a) Λ is closed and bounded, b) Λ contains no intervals, and c) every point in Λ is an accumulation point of Λ.

a) Since Λ is the intersection of closed sets, Proposition 3.15 implies that it is closed. As Λ is contained in $[0,1]$, it is also bounded.

b) If Λ contains the open interval (x,y) with length $|x-y|$, then for each n, (x,y) must be contained in one of the intervals of Λ_n. However, Lemma 8.6 implies that there is $\lambda > 1$ such that the length of an interval in Λ_n is less than $(\frac{1}{\lambda})^n$. Since we can find n_0 such that $|x-y| > (\frac{1}{\lambda})^{n_0}$, the interval (x,y) cannot possibly fit into an interval in Λ_{n_0}. Hence, Λ contains no open intervals.

c) Finally, suppose x is a point in Λ and let $N_\delta(x) = (x-\delta, x+\delta)$ be a neighborhood of x. We must show that $N_\delta(x)$ contains a point in Λ other than x. Notice that if a is an endpoint of one of the intervals in Λ_n, then a is in Λ since $h^{n+1}(a) = 0$. Now for each n, x must be contained in one of the intervals of Λ_n. We let λ be as in Lemma 8.6 and choose n large enough so that $(\frac{1}{\lambda})^n < \delta$. Then the entire interval of Λ_n must be in $N_\delta(x)$ since the length of each interval in Λ_n is less than $(\frac{1}{\lambda})^n$. Since both of the endpoints of the interval are in $N_\delta(x)$ and at least one of them is not x, we are done. □

While we now have a name for Λ (a Cantor set), we still know very little about the dynamics of $h(x)$ on Λ. We begin our investigation of the dynamics by noting that points beginning close to but outside of Λ are in the stable set of infinity and so move away from Λ under iteration of h. Sets with this property are common in dynamics, and many of them are classified as hyperbolic repelling sets.

DEFINITION 8.8. *The set Ω in the domain of the function f is a hyperbolic repelling set if Ω is closed and bounded, $f(\Omega) = \Omega$, and there is $N > 0$ such that $|(f^n)'(x)| > 1$ for all x in Ω and all $n \geq N$. Similarly, the set Ω is a hyperbolic attracting set of the function f if Ω is closed and bounded, $f(\Omega) = \Omega$, and there is $N > 0$ such that $|(f^n)'(x)| < 1$ for all x in Ω and all $n \geq N$.*

In the exercises, we ask the reader to show that Λ is a hyperbolic repelling set of $h_r(x) = rx(1-x)$ when $r > 2 + \sqrt{5}$. Another example of a hyperbolic repelling set is the orbit of a repelling periodic point. Similarly, the orbit of an attracting periodic point is a hyperbolic attracting set. In general, points located near but not in a hyperbolic repelling set move away from the set under iteration of the function. Points located near a hyperbolic attracting set are in the stable set of the hyperbolic set. Here we are generalizing the concept of stable set by saying that x is in the stable set of Ω if the distance from $f^n(x)$ to Ω approaches 0 as n increases. Since a hyperbolic repelling set is closed and bounded (and thus compact), we can consider the distance from a point to the set to be the smallest distance from the point to any point in the set. That is, the distance from x to Ω is the smallest difference $|y - x|$ achievable by allowing y to be any point in Ω. For a more general discussion of the distance from a point to a set, see the texts by Rudin or Willard, which are listed in the references. If we denote the distance from $f^n(x)$ to Ω by $d[f^n(x), \Omega]$, then we can denote the stable set of Ω by

$$W^s(\Omega) = \{x \text{ where } \lim_{n \to \infty} d[f^n(x), \Omega] = 0\}.$$

We summarize the behavior of points near hyperbolic sets in the following theorem.

THEOREM 8.9. *Let Ω be a hyperbolic repelling set of the function f. Then there is an open set U containing Ω such that if x is in U and x is not in Ω, then there is n such that $f^n(x)$ is not in U.*

If Ω is a hyperbolic attracting set of the function f, then there is an open set U containing Ω such that if x is in U, then x is in the stable set of Ω.

The proof of Theorem 8.9 is similar to that of Theorem 6.5. It does require a few facts from introductory real analysis that we haven't developed here. Readers with a background in analysis are encouraged to reread the proof of Theorem 6.5 and adapt it to prove Theorem 8.9.

We return to our consideration of the logistic function by asking: If we restrict the domain of $h(x)$ to Λ, then what can we say about the dynamics of h on Λ? (We note that this question makes sense since h maps Λ into Λ, so we can consider the orbits of points in Λ.) Clearly Λ is a strange set. We know it contains the endpoints of every interval in Λ_n, since these points are eventually fixed at 0. We also know that it contains points other than endpoints since we will show in Exercise 8.4 that every interval of Λ_n contains a periodic point of h with period n. Periodic points aren't eventually fixed, so these periodic points can't be the endpoint of a constituent interval in Λ_m for any m. This is perplexing given that Λ doesn't contain any intervals, and we get from Λ_n to Λ_{n+1} by removing open intervals from the middle of each interval in Λ_n. Our intuition suggests that all that should be left is endpoints of intervals, but this is clearly not the case.

So what can we say about the dynamics of h on Λ? In Exercise 8.4, we will show that the periodic points of Λ are dense in Λ. That is, we will

8.3. Chaos and the Dynamics of the Logistic Function

show that if x is any point in Λ and $\epsilon > 0$, then there is a periodic point p in Λ such that $|x - p| < \epsilon$. In Proposition 8.11, we will show that Λ is "well-mixed" by h or, more precisely, that h is topologically transitive on Λ.

DEFINITION 8.10. *The function $f : D \to D$ is topologically transitive on D if for any open sets U and V that intersect D there is z in $U \cap D$ and a natural number n such that $f^n(z)$ is in V. Equivalently, f is topologically transitive on D if for any two points x and y in D and any $\epsilon > 0$, there is z in D such that $|z - x| < \epsilon$ and $|f^n(z) - y| < \epsilon$ for some n.*

PROPOSITION 8.11. *If $r > 2+\sqrt{5}$, then $h_r(x) = rx(1-x)$ is topologically transitive on Λ, where $\Lambda = \{x | h^n(x) \text{ is in } [0,1] \text{ for all } n\}$.*

PROOF. Let x and y be elements of Λ and $\epsilon > 0$. It suffices to show that there is z in Λ satisfying $|x - z| < \epsilon$ and $h^n(z) = y$ for some n. From Lemma 8.6 we know that there is $\lambda > 1$ such that the length of any interval in Λ_n is less than $(\frac{1}{\lambda})^n$. Choose n so that $(\frac{1}{\lambda})^n < \epsilon$. Since x is in Λ, we know from Proposition 8.1 that there is an interval I_n in Λ_n such that x is in I_n and $h^n : I_n \to [0,1]$ is one-to-one and onto. By our choice of n, the length of I_n is less than ϵ, and so the distance from x, which is a point in I_n, to any point in I_n is less than ϵ. Since $h^n : I_n \to [0,1]$ is one-to-one and onto and y is in $[0,1]$, there is z in I_n such that $h^n(z) = y$. Clearly, z must be in Λ, so we have found z in Λ such that $|x - z| < \epsilon$ and $h^n(z) = y$. □

So far, we have seen that the dynamics of h on Λ contains a certain amount of regularity (the periodic points of h are dense in Λ), and we know that Λ is well-mixed by h. In Proposition 8.13, we show that h also exhibits a certain amount of unpredictability. This unpredictability is called sensitive dependence on initial conditions and is one of the hallmarks of chaos.

DEFINITION 8.12. *The function $f : D \to D$ exhibits sensitive dependence on initial conditions if there exists a $\delta > 0$ such that for any x in D and any $\epsilon > 0$, there is a y in D and a natural number n such that $|x-y| < \epsilon$ and $|f^n(x), f^n(y)| > \delta$.*

Practically speaking, sensitive dependence implies that if we are using an iterated function to model long-term behavior (such as population growth, weather, or economic performance) and the function exhibits sensitive dependence, then any error in measurement of the initial conditions may result in large differences between the predicted behavior and the actual behavior of the system we are modeling. Since all physical measurements include error, this condition may severely limit the utility of our model.

PROPOSITION 8.13. *Let $h_r(x) = rx(1-x)$ and*
$$\Lambda = \{x | h^n(x) \text{ is in } [0,1] \text{ for all } n\}.$$
If $r > 2 + \sqrt{5}$, then $h : \Lambda \to \Lambda$ exhibits sensitive dependence on initial conditions. More precisely, for all x in Λ and all $\epsilon > 0$ there is y in Λ satisfying $|x - y| < \epsilon$ and $|h^n(x) - h^n(y)| > \frac{1}{2}$ for some n.

PROOF. Let x be in Λ and $\epsilon > 0$. Using the same reasoning as in the proof of Proposition 8.11, we see that there is an interval I_n in Λ_n such that x is in I_n, the distance from x to any point in I_n is less than ϵ, and the function $h^n : I_n \to [0,1]$ is one-to-one and onto. Thus, there are points a and b in I_n such that $h^n(a) = 0$ and $h^n(b) = 1$. The points a and b are eventually fixed at 0 so they are in Λ. Since $h^n(x)$ is in Λ and $\frac{1}{2}$ is not in Λ, $h^n(x)$ is in $[0, \frac{1}{2})$ or $h^n(x)$ is in $(\frac{1}{2}, 1]$. If $h^n(x)$ is in $[0, \frac{1}{2})$, then b is in Λ, $|x - b| < \epsilon$, and $|h^n(x) - h^n(b)| = |h^n(x) - 1| > \frac{1}{2}$. If $h^n(x)$ is in $(\frac{1}{2}, 1]$, then a is in Λ, $|x - a| < \epsilon$, and $|h^n(x) - h^n(a)| = |h^n(x)| > \frac{1}{2}$. In either case, we are done. □

Functions exhibiting the three characteristics we have outlined for h on Λ—density of periodic points, topological transitivity, and sensitive dependence—are said to be chaotic.

DEFINITION 8.14. (DEVANEY[1]) *The function $f : D \to D$ is chaotic if*

a) *the periodic points of f are dense in D,*

b) *f is topologically transitive, and*

c) *f exhibits sensitive dependence on initial conditions.*

Restated, this definition implies that a chaotic function observes a certain amount of regularity and mixes the domain well. Further, even the smallest changes in initial position may result in dramatically different results in values under iteration of the function. The regularity is evidenced by the fact that we can always find a periodic orbit in every neighborhood, no matter how small, of every point in the domain of the function. On the other hand, the domain is well-mixed by the function since if we choose any open set, then we can find a point in every other open set that will eventually end up in the first set under iteration of the function. As we have seen, $h : \Lambda \to \Lambda$ exhibits all of these behaviors and is chaotic.

Before leaving this discussion, we note two more facts pertaining to the study of chaotic functions.

---

[1] There are several definitions of chaotic. This definition was introduced by R. Devaney in his text, *Introduction to Chaotic Dynamical Systems*, which was published in 1989 and is listed in the references. Other definitions of chaos are discussed in Section 3.5 of the text *Dynamical Systems: Stability, Symbolic Dynamics, and Chaos* by C. Robinson, which is also listed in the references.

8.3. Chaos and the Dynamics of the Logistic Function

PROPOSITION 8.15. *Let $f : D \to D$. If f is topologically transitive on D, then either D is an infinite set or D consists of the orbit of a single periodic point.*

PROOF. Suppose that D contains only finite many points and $f : D \to D$ is topologically transitive. Choose any point in D and designate it as x_0. We will show that x_0 is periodic and every point in D is in the orbit of x_0. Let $\epsilon > 0$ be such that the distance between any two points in D is at least ϵ. That is, if z and y are in D and $z \neq y$, then $|z - y| \geq \epsilon$. Such an ϵ exists since D is finite. Now let y be any element of D. Since f is topologically transitive, there is z in D and a natural number m such that $|x_0 - z| < \epsilon$ and $|f^m(z) - y| < \epsilon$. But ϵ is the minimum distance between distinct points in D so $|x_0 - z| < \epsilon$ implies that $z = x_0$ and $|f^m(z) - y| < \epsilon$ implies that $f^m(z) = y$. Together, these statements imply that $f^m(x_0) = f^m(z) = y$. Hence, every point in D is in the orbit of x_0. Now choose y_0 to be a point in D other than x_0. By an argument similar to the preceding one, we can show that there is n such that $f^n(y_0) = x_0$. Since we already know that there is m such that $f^m(x_0) = y_0$, it is clear that $f^{n+m}(x_0) = f^n(y_0) = x_0$, and thus x_0 is periodic with period $n + m$.

By contraposition, if D does not consist of the orbit of a single periodic point and f is topologically transitive on D, then D is an infinite set. □

In the following result, which was proven by Banks, Brooks, Cairns, Davis, and Stacey, and appeared in the April 1992 issue of the *American Mathematical Monthly*, we demonstrate that for continuous functions, the defining characteristics of chaos are topological transitivity and the density of periodic points. There is a further assumption that the domain of the function is infinite, but as we demonstrated in Proposition 8.15, if the domain of a topological transitive function is finite, then the dynamics are relatively simple.

THEOREM 8.16. *Let D be an infinite subset of the real numbers and $f : D \to D$ be continuous. If f is topologically transitive on D and the periodic points of f are dense in D, then f is chaotic on D.*

PROOF. It suffices to prove that f exhibits sensitive dependence on initial conditions. We begin by showing that there exists $\delta > 0$ such that for each x in D there is a periodic point q such that $|x - f^i(q)| \geq 4\delta$ for all i.

First, note that a dense subset of an infinite set must be infinite. (The reader should be able to prove this.) Since the periodic points of f are dense in D, this implies that there are infinitely many periodic points and we can choose two, q_1 and q_2, so that $f^n(q_1) \neq f^m(q_2)$ for any n or m. That is, q_1 and q_2 are periodic points that are not in the same orbit. Let δ be the largest number satisfying $\delta \leq \frac{1}{8}|f^n(q_1) - f^m(q_2)|$ for all m and n. Clearly $\delta > 0$, since q_1 and q_2 are periodic and there are only finitely many points of the form $f^n(q_1)$ and $f^m(q_2)$. Thus, there are only finitely many distances of the form $|f^n(q_1) - f^m(q_2)|$, and so there is a smallest such distance. Now if x is any point in D, then the Triangle Inequality implies

$$\delta \leq \tfrac{1}{8}|f^n(q_1) - f^m(q_2)| \leq \tfrac{1}{8}(|f^n(q_1) - x| + |x - f^m(q_2)|)$$

or

$$8\delta \leq |f^n(q_1) - x| + |x - f^m(q_2)|$$

for all m and n. But this implies that either $|x - f^n(q_1)| \geq 4\delta$ for all n or $|x - f^m(q_2)| \geq 4\delta$ for all m. Thus, there is a periodic point q such that $|x - f^i(q)| \geq 4\delta$ for all i.

Now fix a $\delta > 0$ meeting the criterion just described. That is, choose $\delta > 0$ so that for each point in D there is a periodic point that remains at least a distance 4δ away from that point under iteration of f. Let x be an arbitrary point in D and ϵ be such that $0 < \epsilon < \delta$. Fix a periodic point q satisfying $|x - f^i(q)| \geq 4\delta$ for all i. In what follows, we use q to prove that there is w in D satisfying $|w - x| < \epsilon$ and $|f^n(x) - f^n(w)| > \delta$ for some n. As the reader will recall, this implies that f has sensitive dependence on initial conditions. Note that we can make the restriction that $\epsilon < \delta$ since it becomes more difficult to find w as ϵ gets smaller.

We make several observations.

OBSERVATION 1. Since the periodic points of f are dense in D, we can choose one whose distance from x is less than ϵ. We fix one such periodic point and call it p. Let k denote the period of p.

OBSERVATION 2. There is $\mu > 0$ such that $\mu \leq \delta$ and if z is an element of D satisfying $|z - q| < \mu$, then $|f^i(z) - f^i(q)| < \delta$ for all i such that $0 \leq i \leq k$, where k denotes the period of p from Observation 1. This fact follows from the continuity of f.

OBSERVATION 3. There is y in D and a natural number m, so that $|x - y| < \epsilon$ and $|f^m(y) - q| < \mu$, where μ is as in Observation 2. This is an immediate consequence of the topological transitivity of f. Note that in conjunction with Observation 2, the choice of y and m implies that the quantity $|f^{m+i}(y) - f^i(q)| = |f^i(f^m(y)) - f^i(q)|$ is less than δ for all i between 0 and k.

OBSERVATION 4. Given m as in Observation 3, there is n such that $m \leq n \leq m + k$ and k divides n. In other words, $n = ak$, where a is the smallest integer greater than or equal to $\tfrac{m}{k}$.

Now with m and y as in Observation 3 and n as in Observation 4, our choice of δ and the Triangle Inequality imply that

$$\begin{aligned}4\delta &\leq |x - f^{n-m}(q)| \\ &\leq |x - f^n(p)| + |f^n(p) - f^n(y)| + |f^n(y) - f^{n-m}(q)|.\end{aligned}$$

Using the fact that p is periodic with period k and that k divides n, we get

$$4\delta \leq |x - p| + |f^n(p) - f^n(y)| + |f^{n-m}(f^m(y)) - f^{n-m}(q)|.$$

We know from our choice of p in Observation 1 that $|x - p| < \delta$. Also, n was chosen in Observation 4 so that $n - m$ is between 0 and k. Thus, we know from Observation 3 that $|f^{n-m}(f^m(y)) - f^{n-m}(q)| < \delta$ and the previous displayed inequality becomes

$$4\delta \leq \delta + |f^n(p) - f^n(y)| + \delta$$

or

$$2\delta \leq |f^n(p) - f^n(y)|.$$

Using the Triangle Inequality again we get

$$2\delta \leq |f^n(p) - f^n(y)|$$
$$\leq |f^n(p) - f^n(x)| + |f^n(x) - f^n(y)|.$$

But this implies that either $|f^n(p) - f^n(x)| > \delta$ or $|f^n(x) - f^n(y)| > \delta$. Since p and y were chosen in Observations 1 and 3 so that $|p - x| < \epsilon$ and $|y - x| < \epsilon$, we have shown that there is a point w in D (either p or y) satisfying $|w - x| < \epsilon$ and $|f^n(x) - f^n(w)| > \delta$ for some n. □

To summarize the work of this chapter, we collect the known results for the logistic function $h_r(x) = rx(1-x)$ when $r > 4$ in the following theorem.

THEOREM 8.17. *Let* $h_r(x) = rx(1-x)$, $r > 4$, *and*

$$\Lambda = \{x | h^n(x) \text{ is in } [0, 1] \text{ for all } n\}.$$

Then

a) *real numbers not in Λ are in the stable set of infinity,*

b) *the set Λ is a Cantor set, and*

c) *the function $h : \Lambda \to \Lambda$ is chaotic.*

We have proven Theorem 8.17 for the case $r > 2 + \sqrt{5}$. The proof for the case when $4 < r \leq 2 + \sqrt{5}$ is considerably harder and is beyond the scope of this text. Interested readers can find descriptions of several approaches to this problem in the book, *Dynamical Systems: Stability, Symbolic Dynamics, and Chaos* by C. Robinson, which is listed in the references.

8.4. A Few Additional Comments on Cantor Sets

Cantor sets were discovered by the German mathematician Georg Cantor in the last part of the nineteenth century. At the time of their discovery, they created a considerable stir in the mathematical community since their properties are counterintuitive. For example, it is hard to imagine points in the Cantor set Γ other than the endpoints of the intervals in each Γ_n, yet we know that they must exist. Throughout much of the twentieth century, Cantor sets were considered to be little more than a mathematical curiosity. However, the work of Stephen Smale in the 1960s demonstrated that Cantor sets play an important role in dynamical systems. We saw one indication of their importance in Theorem 8.7.

A Cantor set is also an example of a fractal. One of the commonly cited properties of fractals is that they are self-similar under magnification. In the exercises, we ask the reader to show that the Cantor middle-thirds set is homeomorphic to a subset of itself. More generally, we ask the reader to show that if I is an open interval whose intersection with the Cantor middle-thirds set is not empty, then the intersection contains a subset that is homeomorphic to the Cantor middle-thirds set.

Exercise Set 8

8.1 Show that $h(x) = rx(1-x)$ has infinitely many eventually fixed points when $r \geq 4$.

Hint: Consider Proposition 8.1.

8.2 a) Prove that Λ_n is contained in $[0, 1]$ for all n.

b) Explain why $\Lambda_{k+1} \subset \Lambda_k$.

•8.3 Prove Lemma 8.6 with the following steps:

a) Prove that $h'(\frac{1}{2} - \frac{\sqrt{r^2-4r}}{2r}) > 1$ and $h'(\frac{1}{2} + \frac{\sqrt{r^2-4r}}{2r}) < -1$ when $r > 2 + \sqrt{5}$. Show that $h''(x) < 0$ for all x and conclude that $h'(x)$ is decreasing. Explain why these two statements guarantee that for each r there is $\lambda > 1$ so that $|h'(x)| > \lambda$ when x is in Λ_1.

b) Use part (a) and Exercise 6.2 to show that if c is any point in Λ_n, then $(h^n)'(c) > \lambda^n$.

c) Use the Mean Value Theorem and part (b) to show that if a and b are the endpoints of one of the constituent intervals of Λ_n, then $|h^n(b) - h^n(a)| > \lambda^n |b - a|$. Explain how we know that

$|h^n(b) - h^n(a)| = 1$, and conclude that $|b - a|$, which is the length of the interval in Λ_n, is less than $(\frac{1}{\lambda})^n$.

8.4 a) Use Proposition 8.1 and Theorem 4.4 to show that if $r > 4$, then $h_r(x) = rx(1-x)$ has 2^n periodic points with period n.

b) Show that if $r > 2 + \sqrt{5}$, then the periodic points found in part (a) are repelling. You may wish to use Exercise 8.3b.

c) Use Lemma 8.6 to show that if $r > 2 + \sqrt{5}$, then the periodic points of h_r are dense in Λ.

8.5 Suppose $m > 1$ and let $f : \mathbb{R} \to \mathbb{R}$ be defined by $f(x) = mx$. Show that f exhibits sensitive dependence on initial conditions. Is f chaotic? Explain.

8.6 a) Let $g : \mathbb{R} \to \mathbb{R}$ be defined by $g(x) = \begin{cases} 3x, & \text{for } x \leq \frac{1}{2} \\ 3 - 3x, & \text{for } x \geq \frac{1}{2}. \end{cases}$ Show that the set

$$\hat{\Gamma} = \{x \text{ in } [0,1] \mid g^n(x) \text{ is in } [0,1] \text{ for all } n\}$$

is the Cantor middle-thirds set.

b) Let Γ be the Cantor middle-thirds set. Use part (a) to demonstrate that the map $f : \Gamma \to [0, \frac{1}{3}] \cap \Gamma$ defined by $f(x) = \frac{1}{3}x$ is a homeomorphism of Γ to a subset of itself.

c) Show that if I is one of the intervals in Γ_n from the construction of the Cantor middle-thirds set (see Example 8.3), then Γ is homeomorphic to $\Gamma \cap I$.

d) Prove that if I is an open interval whose intersection with the Cantor middle-thirds set is not empty, then the intersection of I and the Cantor middle-thirds set contains a subset that is homeomorphic to the Cantor middle-thirds set.

8.7 Show that Λ is a hyperbolic repelling set of $h(x) = rx(1-x)$ when $r > 2 + \sqrt{5}$.

*8.8 Let D be an interval and $f : D \to D$ be a continuous function.

a) Suppose that for any nontrivial closed intervals I and J contained in D, we can find n so that $f^n(I) \supset J$. Prove that f is chaotic on D.

b) Is the converse of part (a) true?

- 8.9 THE TENT MAP:

    Let $T : [0, 1] \to [0, 1]$ be the tent map defined by

    $$T(x) = \begin{cases} 2x, & \text{for } x \text{ in } [0, \tfrac{1}{2}] \\ 2 - 2x, & \text{for } x \text{ in } [\tfrac{1}{2}, 1]. \end{cases}$$

    a) Graph $T(x)$, $T^2(x)$, $T^3(x)$, and $T^4(x)$. (You may wish to use a computer graphing tool. See Section A.1 in the Appendix for suggestions on how to define T for the computer.)

    b) Show that T has 2^n repelling periodic points with period n.

    Hint: It may help to begin by showing that for integers k between 0 and $2^n - 1$ the function $T^n : \left[\tfrac{k}{2^n}, \tfrac{k+1}{2^n}\right] \to [0, 1]$ is an onto linear function.

    c) Show that the periodic points of T are dense in $[0, 1]$.

    d) Show that T is topologically transitive on $[0, 1]$.

    e) Prove that $T : [0, 1] \to [0, 1]$ exhibits sensitive dependence on initial conditions. (This can easily be done without appealing to Theorem 8.16.)

    f) Conclude that T is chaotic on $[0, 1]$.

8.10 Show that the two descriptions of topological transitivity given in Definition 8.10 are equivalent.

*8.11 Prove Theorem 8.9.

8.12 A COMPUTER INVESTIGATION: Use a computer program that allows the user to control the the amount of rounding in its calculations to empirically test whether or not $h : [0, 1] \to [0, 1]$ defined by $h(x) = 4x(1 - x)$ exhibits sensitive dependence on initial conditions. An example of such a *Mathematica* program can be found in the Appendix.

**8.13 Prove that $h(x) = rx(1 - x)$ is chaotic on Λ when $4 < r \leq 2 + \sqrt{5}$.

9

The Logistic Function Part II: Topological Conjugacy

We continue our investigation of the logistic function by showing that $h(x) = 4x(1-x)$ is chaotic on $[0, 1]$. Unfortunately, proving this directly from the definition is a relatively difficult task. Consequently, we will show instead that the dynamics of h on $[0, 1]$ are the same as the dynamics of the tent map on $[0, 1]$. Mathematically speaking, we say that h on $[0, 1]$ is topologically conjugate to the tent map on $[0, 1]$. (The tent map was introduced and shown to be chaotic in Exercise 8.9 on page 86.)

DEFINITION 9.1. *Let $f : D \to D$ and $g : E \to E$ be functions. Then f is topologically conjugate to g if there is a homeomorphism $\tau : D \to E$ such that $\tau \circ f = g \circ \tau$. In this case, τ is called a topological conjugacy.*

We represent this relationship by the commutative diagram

$$\begin{array}{ccc} D & \xrightarrow{f} & D \\ \tau \downarrow & & \downarrow \tau \\ E & \xrightarrow{g} & E. \end{array}$$

When we say that the diagram is commutative, we mean that if we start with an element in D and follow the arrows to another set in the diagram, we arrive at the same element no matter which route we take. For example, let x be an element of D in the upper left corner of the diagram. Follow the arrows from the upper left corner over to get $f(x)$ in D and then down to get $\tau(f(x))$ in E. Going in the other direction, follow the arrow down to get $\tau(x)$ in E and then over to get $g(\tau(x))$. Since the diagram is commutative, we have arrived at the same element by both paths and so we know that $\tau(f(x)) = g(\tau(x))$. We can represent this by another, more detailed commutative diagram:

$$\begin{array}{ccc} x \text{ in } D & \xrightarrow{f} & f(x) \text{ in } D \\ \tau\downarrow & & \downarrow\tau \\ \tau(x) \text{ in } E & \xrightarrow{g} & \tau(f(x)) = g(\tau(x)) \\ & & \text{in } E \end{array}$$

Now, τ is a homeomorphism, so we know that τ^{-1} is defined and

$$\tau^{-1}(\tau(f(x))) = f(x).$$

Since τ commutes with f and g, we can substitute $g(\tau(x))$ for $\tau(f(x))$ in the previous equation to get

$$\tau^{-1}(g(\tau(x))) = f(x).$$

We represent this in a commutative diagram by

$$\begin{array}{ccc} D & \xrightarrow{f} & D \\ \tau\downarrow & & \uparrow\tau^{-1} \\ E & \xrightarrow{g} & E \end{array} \quad \text{or} \quad \begin{array}{ccc} x & \xrightarrow{f} & f(x) = \tau^{-1}(g(\tau(x))) \\ \tau\downarrow & & \uparrow\tau^{-1} \\ \tau(x) & \xrightarrow{g} & g(\tau(x)). \end{array}$$

Similarly, if e is an element of E, then $\tau(f(\tau^{-1}(e))) = g(e)$ or

$$\begin{array}{ccc} D & \xrightarrow{f} & D \\ \tau^{-1}\uparrow & & \downarrow\tau \\ E & \xrightarrow{g} & E \end{array} \quad \text{or} \quad \begin{array}{ccc} \tau^{-1}(e) & \xrightarrow{f} & f(\tau^{-1}(e)) \\ \tau^{-1}\uparrow & & \downarrow\tau \\ e & \xrightarrow{g} & g(e). \end{array}$$

Returning now to the definition of a homeomorphism, we recall that $\varphi : D \to E$ is a homeomorphism if and only if φ is continuous, one-to-one, onto, and has a continuous inverse. Further, we note that if D and E are homeomorphic (i.e., there exists a homeomorphism from D to E), then the topologies of D and E are the same. We make this precise in the following proposition.

9. The Logistic Function Part II: Topological Conjugacy

PROPOSITION 9.2. *Let D and E be subsets of the real numbers and the function $\varphi : D \to E$ be a homeomorphism. Then*

a) *the set U in D is open if and only if the set $\varphi(U)$ is open in E,*

b) *the sequence $x_1, x_2, x_3, \ldots$ in D converges to x in D if and only if the sequence $\varphi(x_1), \varphi(x_2), \varphi(x_3), \ldots$ converges to $\varphi(x)$ in E,*

c) *the set F is closed in D if and only if the set $\varphi(F)$ is closed in E, and*

d) *the set A is dense in D if and only if the set $\varphi(A)$ is dense in E.*

The proof of this proposition follows immediately from the relevant definitions and is left to the reader. Readers interested in proving this result may find it useful to review the material in Chapter 3.

When functions of D and E are topologically conjugate, the existence of the homeomorphism from D to E guarantees that the topologies of the two spaces are identical. The condition that $\tau \circ f = g \circ \tau$ guarantees that the dynamics are the same. This is demonstrated in Theorem 9.3.

THEOREM 9.3. *Let D and E be subsets of the real numbers, $f : D \to D$, $g : E \to E$, and $\tau : D \to E$ be a topological conjugacy of f and g. Then*

a) $\tau^{-1} : E \to D$ *is a topological conjugacy,*

b) $\tau \circ f^n = g^n \circ \tau$ *for all natural numbers n, and*

c) *p is a periodic point of f if and only if $\tau(p)$ is a periodic point of g. Further, the prime periods of p and $\tau(p)$ are identical.*

d) *If p is a periodic point of f with stable set $W^s(p)$, then the stable set of $\tau(p)$ is $\tau(W^s(p))$.*

e) *The periodic points of f are dense in D if and only if the periodic points of g are dense in E.*

f) *f is topologically transitive on D if and only if g is topologically transitive on E.*

g) *f is chaotic on D if and only if g is chaotic on E.*

Before we begin the proof of Theorem 9.3 it is worth noting that many students of mathematics (both professional and those with amateur status) find it useful to use commutative diagrams to keep track of arguments involving topological conjugacies.

Let us consider, for example, the proof of part (c) of this theorem. Suppose p is a periodic point of f with prime period k. We wish to show that $\tau(p)$ is a periodic point of g with prime period k. To help us develop the proof, we consider the diagram

$$
\begin{array}{ccc}
p \text{ in } D & \xrightarrow{f^k} & f^k(p) = p \text{ in } D \\
\tau \downarrow & & \downarrow \tau \\
\tau(p) \text{ in } E & \xrightarrow{g^k} & g^k(\tau(p)) = \tau(f^k(p)).
\end{array}
$$

From part (b), we know that this diagram commutes, so

$$g^k(\tau(p)) = \tau(f^k(p)) = \tau(p)$$

and we have shown that $\tau(p)$ is a periodic point of g with period k. It remains to show that $\tau(p)$ has prime period k. If $0 < n < k$, then we see that $g^n(\tau(p)) = \tau(f^n(p)) \neq \tau(p)$ since τ is one-to-one and $f^n(p) \neq p$ if $0 < n < k$. Hence, $\tau(p)$ is a periodic point of g with prime period k.

With a little practice, this diagram chasing becomes second nature and an easy way to complete proofs involving commuting functions like topological conjugacies.

PROOF OF THEOREM 9.3.

a) It suffices to show $\tau^{-1} \circ g = f \circ \tau^{-1}$. The details are left to the reader.

b) To prove that $\tau \circ f^n = g^n \circ \tau$, we need only consider the following commutative diagram consisting of n squares. As each of the squares commutes, the whole diagram commutes. Thus, we get the same result by going around the outside of the diagram in either direction, and the truth of the assertion follows.

$$
\begin{array}{ccccccccc}
D & \xrightarrow{f} & D & \xrightarrow{f} & D & \xrightarrow{f} & \cdots & \xrightarrow{f} & D \\
\tau \downarrow & & \downarrow \tau & & \downarrow \tau & & & & \downarrow \tau \\
E & \xrightarrow{g} & E & \xrightarrow{g} & E & \xrightarrow{g} & \cdots & \xrightarrow{g} & E
\end{array}
$$

Alternatively, for those with a more analytical bent, we may use the principle of mathematical induction. The statement $\tau \circ f^n = g^n \circ \tau$ is true when $n = 1$ since τ is a topological conjugacy. Now, suppose that it is true for n, that is, $\tau \circ f^n = g^n \circ \tau$. Then it is true for $n + 1$ since

$$\tau \circ f^{n+1} = (\tau \circ f^n) \circ f = (g^n \circ \tau) \circ f = g^n \circ (\tau \circ f) = g^n \circ (g \circ \tau) = g^{n+1} \circ \tau.$$

c) The fact that τ maps periodic points of f to periodic points of g was shown in the discussion preceding this proof. Since τ^{-1} is a topological conjugacy, the same argument demonstrates that if q is a periodic point of g with prime period k, then $\tau^{-1}(q)$ is a periodic point of f with prime period k.

9. The Logistic Function Part II: Topological Conjugacy 91

d) Let p be a periodic point of f with prime period k and let x be an element of $W^s(p)$. (It may help to keep track of the argument using a commutative diagram.) Then, by definition of $W^s(p)$, for each $\delta > 0$ there is N such that if $n \geq N$, then $|f^{kn}(x) - p| < \delta$. We must show that for each $\epsilon > 0$, there is M such that whenever $n \geq M$, then $|g^{kn}(\tau(x)) - \tau(p)| < \epsilon$.

Suppose $\epsilon > 0$. Since τ is continuous, there exists $\delta > 0$ such that if $|y - p| < \delta$, then $|\tau(y) - \tau(p)| < \epsilon$. Choose M so that if $n \geq M$, then $|f^{kn}(x) - p| < \delta$. Then, by continuity,

$$|\tau(f^{kn}(x)) - \tau(p)| = |g^{kn}(\tau(x)) - \tau(p)| < \epsilon$$

when $n \geq M$ and the proof of (d) is complete.

e) The preservation of density of periodic points by a topological conjugacy follows easily from part (c) of this proposition and part (d) of Proposition 9.2.

f) Let $f: D \to D$ be topologically transitive. We wish to show that $g: E \to E$ is topologically transitive. It suffices to show that if x and y are in E and $\epsilon > 0$, then there is z in E such that $|z - x| < \epsilon$ and $|g^n(z) - y| < \epsilon$ for some n. Consider the elements of D, $x' = \tau^{-1}(x)$ and $y' = \tau^{-1}(y)$. Since τ is continuous, there is δ_x such that

$$\text{if } |w - x'| < \delta_x, \text{ then } |\tau(w) - \tau(x')| = |\tau(w) - x| < \epsilon. \qquad (9.1)$$

Similarly, there is δ_y such that

$$\text{if } |w - y'| < \delta_y, \text{ then } |\tau(w) - \tau(y')| = |\tau(w) - y| < \epsilon. \qquad (9.2)$$

Let δ be the smaller of δ_x and δ_y. Then, since f is topologically transitive, there is z' in D such that $|x' - z'| < \delta$ and $|f^n(z') - y'| < \delta$ for some n. Now let $z = \tau(z')$. Since $|x' - z'| < \delta \leq \delta_x$, condition (9.1) implies that $|z - x| = |\tau(z') - \tau(x')| < \epsilon$. Similarly, condition (9.2) implies that $|g^n(z) - y| = |\tau(f^n(z')) - \tau(y')| < \epsilon$ since $|f^n(z') - y'| < \delta \leq \delta_y$. Consequently, g is topologically transitive on E.

Since τ^{-1} is a topological conjugacy from g to f, a similar argument will show that the converse is true as well.

g) If f and g are continuous, then this follows from parts (e) and (f) and Theorem 8.16. Since the general case requires more tools and is not needed by us, we leave that as an exercise for the truly inspired reader. □

Before we move on, we note that in general sensitive dependence is *not* preserved by topological conjugacy. However, if the domains of f and g are closed and bounded (or in more general settings, compact), then sensitive dependence is preserved by topological conjugacy. For details, the reader may consult Section 3.5 of the text *Dynamical Systems: Stability, Symbolic Dynamics, and Chaos* by C. Robinson.

9. The Logistic Function Part II: Topological Conjugacy

To demonstrate that sensitive dependence is not always preserved by topological conjugacies, consider the function $f : (0,1) \to (0,1)$ defined by $f(x) = x^2$ and the function $g : (1,\infty) \to (1,\infty)$ defined by $g(x) = x^2$. We leave it to the reader to show that $f(x)$ does not exhibit sensitive dependence on initial conditions, even though $\tau(x) = \frac{1}{x}$ is a topological conjugacy between $f(x)$ and $g(x)$, and $g(x)$ does exhibit sensitive dependence on initial conditions. For more details, see Exercise 9.3.

We are now ready to use a topological conjugacy to show that the function $h(x) = 4x(1-x)$ is chaotic on $[0,1]$.

THEOREM 9.4. *The logistic function $h_4 : [0,1] \to [0,1]$ defined by $h(x) = 4x(1-x)$ is topologically conjugate to the tent map $T : [0,1] \to [0,1]$ defined by*

$$T(x) = \begin{cases} 2x, & \text{for } x \text{ in } [0, \tfrac{1}{2}] \\ 2 - 2x, & \text{for } x \text{ in } [\tfrac{1}{2}, 1]. \end{cases}$$

PROOF. It suffices to show that the map $\tau(x) = \sin^2(\frac{\pi}{2}x)$ is a topological conjugacy between the two maps. In particular, we must show that

a) $\tau : [0,1] \to [0,1]$ is one-to-one and onto,
b) $\tau : [0,1] \to [0,1]$ is continuous,
c) $\tau^{-1} : [0,1] \to [0,1]$ is continuous, and
d) $\tau \circ T = h \circ \tau$.

The first three properties follow from the fact that $\tau'(x)$ exists on $[0,1]$ and is strictly positive on $(0,1)$. The last property follows directly from elementary trigonometric identities. The details are left to the reader. □

COROLLARY 9.5. *The function $h(x) = 4x(1-x)$ is chaotic on $[0,1]$.*

PROOF. This is an immediate consequence of Theorems 9.3 and 9.4 and Exercise 8.9. □

Exercise Set 9

9.1 Prove Proposition 9.2.

9.2 Prove part (a) of Theorem 9.3. That is, let $\tau : D \to E$ be a topological conjugacy of the maps $f : D \to D$ and $g : E \to E$. Show that $\tau^{-1} : E \to D$ is a topological conjugacy of g to f.

9.3 Consider the function $f : (0,1) \to (0,1)$ defined by $f(x) = x^2$ and the function $g : (1,\infty) \to (1,\infty)$ defined by $g(x) = x^2$. Prove that sensitive dependence is not preserved by topological conjugacy by completing the following steps.

a) Prove that $f(x)$ does *not* exhibit sensitive dependence on initial conditions.

b) Prove that $g(x)$ *does* exhibit sensitive dependence on initial conditions.

c) Show that $\tau(x) = \frac{1}{x}$ is a topological conjugacy between $f(x)$ and $g(x)$.

d) Conclude that topological conjugacies do not necessarily preserve sensitive dependence on initial conditions.

9.4 Write a complete proof of Theorem 9.4; a sketch of the proof follows the statement of the theorem.

•9.5 THE QUADRATIC MAP:

a) Let $f(x) = Ax^2 + Bx + C$ be a quadratic function (where $A \neq 0$). Show that there exists a linear function $\tau(x) = mx + d$ and a parameter c such that τ is a topological conjugacy between f and the quadratic map $q_c(x) = x^2 + c$.

Hint: Simplify the compositions $\tau \circ f$ and $q_c \circ \tau$ and equate coefficients of like terms.

b) Show that if $c \leq 1/4$, then $q_c(x) = x^2 + c$ is topologically conjugate to $h(x) = rx(1 - x)$ for some r. Use the topological conjugacy to prove that q_c has a single attracting fixed point when $-3/4 < c < 1/4$. What happens when $c < -3/4$? What are the dynamics of q_c when $c > 1/4$?

Determine parameter values for which q_c is chaotic on a Cantor set. You may assume that h_r is chaotic on a Cantor set whenever $r \geq 4$.

9.6 Show that $q(x) = x^2 - 2$ is chaotic on the interval $[-2, 2]$.
Hint: You may wish to use the previous exercise.

10
The Logistic Function Part III: A Period-Doubling Cascade

In Chapter 7, we characterized the behavior of the parametrized family of functions $h_r(x) = rx(1-x)$ for $0 < r \leq 3$. By this time, the reader should find it easy to verify that h has, at most, two periodic points for these parameter values, both of which are fixed. All other real numbers are in the stable set of one of these points or the stable set of infinity. In Chapter 8 we demonstrated that there is a Cantor set in $[0,1]$ on which h is chaotic when $r > 2 + \sqrt{5}$. All real numbers not in the Cantor set are in the stable set of infinity. We stated without proof that this property also holds for $4 < r \leq 2 + \sqrt{5}$. Finally, we note in Corollary 9.5 that h is chaotic on $[0,1]$ when $r = 4$. The analysis of the behavior of h for parameter values between 3 and 4 remains.

In Example 7.5, we demonstrated that h_r undergoes a period-doubling bifurcation when $r = 3$. The fixed point $p_r = \frac{r-1}{r}$ changes from being an attracting fixed point when $r < 3$ to a repelling fixed point when $r > 3$. The bifurcation is called period-doubling because when the change occurs an attracting period two orbit is added that straddles p_r. We can visualize the attracting fixed point as splitting into a period two attracting orbit with a repelling fixed point lodged between the two points in the orbit.

To analyze what happens as we increase r from 3 to 4, we use a theorem of Pierre Fatou, which was discovered shortly after the turn of the century.

THEOREM 10.1. [FATOU] *If the quadratic polynomial $f(x) = ax^2+bx+c$ has an attracting periodic orbit, then the critical point $\frac{-b}{2a}$ is in the stable set of one of the points in the orbit.*

We recall from calculus that x_0 is a critical point of the differentiable function f if $f'(x_0) = 0$. Note that the theorem does not state that there is a critical point in the stable set of each attracting periodic point. As there is only one critical point, that would be impossible. What it does say is that there is a critical point in the stable set of one of the iterates of an attracting periodic point. As an easy corollary to Theorem 10.1, we see that a quadratic polynomial can have at most one attracting periodic orbit. We will not prove Theorem 10.1 since the proof requires a substantial amount of material from complex analysis. A proof can be found in R. Devaney's text *Introduction to Chaotic Dynamical Systems*.

We note that the critical point of $h_r(x) = rx(1-x)$ is $\frac{1}{2}$. Consequently, if h_r has an attracting orbit for some r, then we know that $\frac{1}{2}$ is in the stable set of the orbit. We exploit this information to draw a bifurcation diagram for h_r. In doing so, the parameter r is graphed on the horizontal axis, and the value of $h_r^n(\frac{1}{2})$ is plotted on the vertical axis for values of n between 100 and 300. For example, when $r = 3.8$, $h_r^{100}(\frac{1}{2}) = .83668$ and $h_r^{101}(\frac{1}{2}) = .51924$, so the points $(3.8, .83668)$ and $(3.8, .51924)$ are plotted. We continue by plotting $(3.8, h_r^{102}(\frac{1}{2}))$, $(3.8, h_r^{103}(\frac{1}{2}))$, ..., $(3.8, h_r^{300}(\frac{1}{2}))$. This is done for 400 r values between 2 and 4; the resulting graph is shown in Figure 10.1. Since an attracting periodic orbit attracts the critical point, the diagram should give us a picture of the location of the attracting orbits.

Note that we get the behavior we expect for $2 < r < 3$. After 100 iterations, the points have more or less converged to the attracting fixed point $p_r = \frac{r-1}{r}$. When $r = 3$, there is a period-doubling bifurcation and the orbit of the critical point is attracted to a period two orbit when r is a little larger than 3. It appears that another period-doubling bifurcation occurs at a point somewhat less than $r = 3.5$. At that point, the period two orbit splits into an attracting period four orbit, followed by another split into period eight, another into period sixteen, and so on. While an analytic proof that this is what happens is beyond the scope of this book, a convincing geometric argument is readily accessible.

Consider again the graphs of h^2 shown in Figure 7.5 on page 65. Note that as the parameter value passes 3, the derivative of h_r^2 at the fixed point passes from being less than 1 to being greater than 1. Recall that the fixed point is $p_r = \frac{r-1}{r}$. When $r > 3$, the fact that $(h_r^2)(p_r) > 1$ implies that there is a point x_0 to the right of the fixed point satisfying $h_r^2(x_0) > x_0$. Since $h_r^2(1) = 0$, the Intermediate Value Theorem guarantees that there is a fixed point of h_r^2 in the interval $(x_0, 1)$. This point is a prime period two point of h. The reader was asked to prove this in Exercise 7.3.

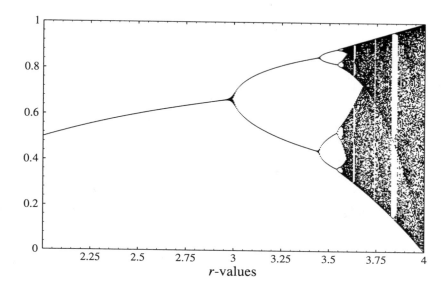

FIGURE 10.1. A bifurcation diagram of $h_r(x) = rx(1-x)$ generated by graphing the 100th through 500th iterates of $\frac{1}{2}$. The parameter value r is plotted on the horizontal axis and the x value is plotted on the vertical axis. The parameter value is allowed to vary from 2 to 4. The range of x values shown is from 0 to 1.

Now consider the graph of $h_r^2(x)$ when $2 < r \leq 4$. For example, consider the graph of $h_{3.4}^2(x)$, which is shown in Figure 10.2. We claim that the graph and thus the dynamics of h_r^2 on $[\frac{1}{r}, \frac{r-1}{r}]$ are qualitatively the same as the graph and the dynamics of $h_s(x)$ on $[0, 1]$ for some s. To see this, we draw a box on the graph of $h_r^2(x)$ starting at the fixed point $(\frac{r-1}{r}, \frac{r-1}{r})$. Drawing a line horizontally to the left, we strike the graph of $h_r^2(x)$ at the point $(\frac{1}{r}, \frac{r-1}{r})$. Turning to draw a line vertically, we strike the graph of $y = x$ at $(\frac{1}{r}, \frac{1}{r})$. Finally, we complete the box by adding a horizontal line segment from $(\frac{1}{r}, \frac{1}{r})$ to $(\frac{r-1}{r}, \frac{1}{r})$ and a vertical line segment from there up to our starting point at $(\frac{r-1}{r}, \frac{r-1}{r})$. The box we have drawn is a square with sides of length $\frac{r-2}{r}$. This box has been drawn on the graph of $h_{3.4}^2(x)$ in Figure 10.2 along with a similar box with corners at $(\frac{r-1}{r}, \frac{r-1}{r})$, $(\frac{r+\sqrt{r^2-4}}{2r}, \frac{r-1}{r})$, $(\frac{r+\sqrt{r^2-4}}{2r}, \frac{r+\sqrt{r^2-4}}{2r})$, and $(\frac{r-1}{r}, \frac{r+\sqrt{r^2-4}}{2r})$. Notice the similarity between the graph of h_s in the unit square when $s = 2.646$ and the graph of $h_{3.4}^2(x)$ in these boxes. It is easiest to see the similarity in the larger box if it is rotated 180°. Given the similarity of the graphs, we expect the dynamics to be similar as well.

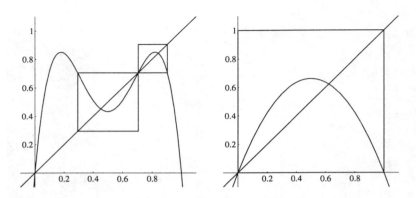

FIGURE 10.2. The graph of $h^2_{3.4}$ and $h_{2.646}$. Notice the similarity of the graphs in the boxed regions.

In Figure 10.3, we see the graphs of $h^2_r(x)$ for a variety of values of r between 2 and 4. On each graph, we have drawn the squares just described. Notice how the graph changes inside the boxes as r increases. The changes in the graph of $h^2_r(x)$ as r increases from 2 are qualitatively the same as the changes in the graph of $h_s(x)$ on $[0, 1]$ as s increases from 0. In particular, we note that when $h^2_r(\frac{1}{2}) = \frac{1}{r}$, the local minimum at $\frac{1}{2}$ is tangent to the bottom of the box. This is qualitatively the same as the graph of $h_4(x)$ on the unit square. Since $h^2_r(\frac{1}{2}) = \frac{r^2(r-4)}{16}$, we can find the r value at which this tangency occurs by solving the equation $\frac{r^2(r-4)}{16} = \frac{1}{r}$. Doing so, we find that the tangency occurs when r is approximately 3.6786. Looking again at Figure 10.3, notice that when r is larger than 3.6786, the graph of $h^2_r(x)$ in the boxes is qualitatively similar to the graph of $h_s(x)$ in the unit square when s is larger than 4. Thus, we expect the dynamics to be similar but there is one important difference. Points that leave the interval $[0, 1]$ under iteration of $h_s(x) = sx(1-x)$ when $s > 4$ are in the stable set of infinity and never return to $[0, 1]$. However, points that leave the interval $[\frac{1}{r}, \frac{r-1}{r}]$ under iteration of $h^2_r(x)$ when $3.6786 \leq r \leq 4$ must stay in the interval $[0, 1]$ and can return to the interval under iteration of $h^2_r(x)$. The return of such a point is illustrated in Figure 10.4. While the potential return of points to $[\frac{1}{r}, \frac{r-1}{r}]$ complicates the dynamics of $h^2_r(x)$, we still expect that there is a Cantor set in $[\frac{1}{r}, \frac{r-1}{r}]$ on which $h^2_r(x)$ behaves chaotically when r is larger than 3.6786. This is indeed true, but the proof is beyond the scope of this text. Similarly, there is another Cantor set in the interval $[\frac{r-1}{r}, \frac{r+\sqrt{r^2-4}}{2r}]$ on which $h^2_r(x)$ is chaotic when r is larger than 3.6786. In addition, points that leave $[\frac{r-1}{r}, \frac{r+\sqrt{r^2-4}}{2r}]$ under iteration of $h^2_r(x)$ may return to the interval after additional iterations. The reader is invited to find such points.

10. The Logistic Function Part III: A Period-Doubling Cascade

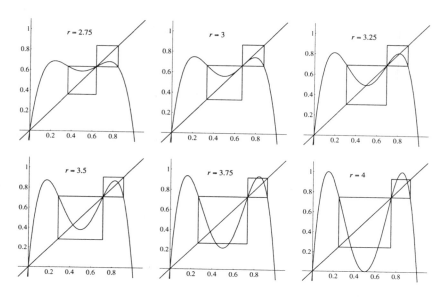

FIGURE 10.3. The graph of h_r^2 for a variety of values of r. Note that for a fixed value of r, the graphs within the boxed regions are qualitatively similar. Also, as r increases from 2 to 4, the graphs within the boxed regions mimic the graphs of $h_s(x)$ on $[0, 1]$ as s increases from 0 to 5.

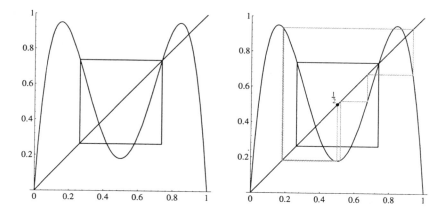

FIGURE 10.4. The graph of h_r^2 when $r = 3.8$. Note that points in $[\frac{1}{r}, \frac{r-1}{r}]$ may leave the interval under iteration of h_r^2 and then return. This is illustrated using graphical analysis beginning at the point $\frac{1}{2}$.

We still haven't described how the dynamics of $h_r(x)$ change as r increases from 3 to 4. In the preceding paragraphs, we argue that $h_r^2(x)$ and hence $h_r(x)$ behave chaotically on some subset of their domain when r is larger than 3.6786, but what happens when r is smaller than that? We recall that in Example 7.6 we saw a period-doubling bifurcation occur when $r = 3$, and an attracting periodic point with period two was formed. In Exercises 10.5, we ask the reader to show that these two attracting periodic points persist as r increases to $1 + \sqrt{6}$ and attract most points in the interval $[0, 1]$. Since $1 + \sqrt{6} \approx 3.45$, we look at the graphs of $h_r^4(x)$ for $r = 3.4$, $r = 3.45$, and $r = 3.5$ in Figure 10.5 to try to determine what happens as r increases beyond $1 + \sqrt{6}$. In each case, the graph is shown on the interval $[\frac{1}{r}, \frac{r-1}{r}]$. Note that the interval changes as we change r, but the behavior is identical to that which we observed when we looked at the graph of $h_r^2(x)$ for r near 3 in Example 7.6; a period-doubling bifurcation occurs. If p is one of the period two points, then $(h_r^4)'(p) < 1$ when $r < 1 + \sqrt{6}$. If $r = 1 + \sqrt{6}$, then $(h_r^4)'(p) = 1$ and the period two point is nonhyperbolic. When $r > 1 + \sqrt{6}$, a new attracting period four point has appeared. In Exercise 10.5, we ask the reader to demonstrate that this is the case.

Looking at the graph of h_r^4 for $r = 3.55$, we again notice square boxes in which the graph of h_r^4 is similar to the graph of h_s in the unit square. These boxes are identified on the graph of $h_{3.55}^4(x)$ in Figure 10.6. Note that the boxes each have a corner at a prime period two point, just like the boxes on the graph of $h_r^2(x)$ each had a corner at a fixed point of $h_r(x)$, that is, a prime period one point. Also, there is one box for each point in the prime period four orbit. As the parameter value increases, the absolute value of the derivative of $h_r^4(x)$ at the period four points exceeds 1 and there is another period-doubling bifurcation; an attracting periodic orbit with period eight is created and the period four orbit becomes repelling. Once the period eight orbit is created, we see eight boxes within which the graph of $h_r^8(x)$ resembles the graph of $h_s(x)$ on the unit square for some s. As the parameter continues to increase, we have another period-doubling bifurcation and an attracting period sixteen orbit is created. This process continues ad infinitum and is called a *period-doubling cascade* or the *period-doubling route to chaos*. One interesting aspect of this cascade is that the ratio of the distances between successive parameter values at which bifurcations occur asymptotically approaches a fixed value called the *Feigenbaum constant*. Since the Feigenbaum constant is approximately 4.669, we can predict at which parameter value the period-doubling cascade will have been completed, and there will be periodic orbits of order 2^n for every n. The Feigenbaum constant is explored in Exercise 10.8.

In general, if h_r has a periodic point with prime period n, then we can find n regions on the graph of h_r^n around which we can draw a square box

10. The Logistic Function Part III: A Period-Doubling Cascade   101

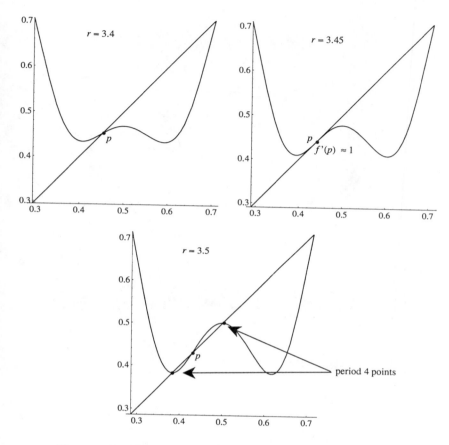

FIGURE 10.5. Graphs of h_r^4 for the parameter values 3.4, 3.45, and 3.5. The point p is a prime period two point of h that undergoes a period-doubling bifurcation when r is approximately 3.45. In each case, the bounds of the graphed region are $\frac{1}{r}$ and $\frac{r-1}{r}$.

so that the graph of h_r^n in the box is similar to the graph of h_s on the unit square. An example of this is shown for a period three point when $r = 3.88$ in Figure 10.7. As r changes, we find all of the same dynamics for h_r^n within these boxes that we find for h_s on the unit square. In particular, since h_s has a period-doubling bifurcation when $s = 3$, h_r^n has a period-doubling bifurcation for some value of r, and a period $2n$ orbit is formed.

As we mentioned earlier, there is a Cantor set in $[\frac{1}{r}, \frac{r-1}{r}]$ on which h_r^2 is chaotic whenever $r \geq 3.6786$. As a result, we expect that the dynamics for

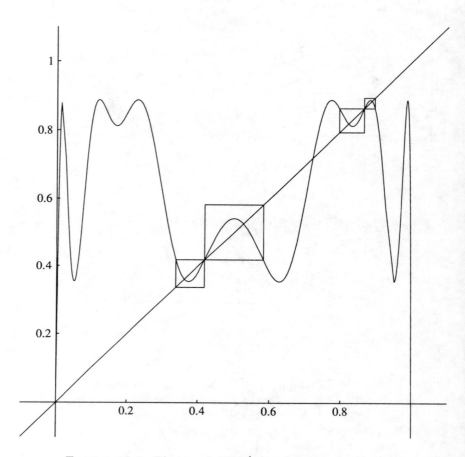

FIGURE 10.6. The graph of h_r^4 for the parameter value $r = 3.55$. Notice that the graph of $h_r^4(x)$ is similar in each of the boxes and is qualitatively the same as the graph of $h_x(x)$ on $[0, 1]$ for some s.

10. The Logistic Function Part III: A Period-Doubling Cascade 103

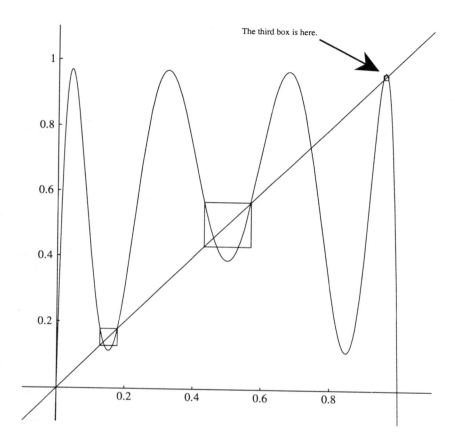

FIGURE 10.7. The graph of h_r^3 for the parameter value $r = 3.88$. Note the boxes in which the graph is similar to that of the logistic function on $[0, 1]$.

parameter values in the interval [3.6786, 4] will be extremely complicated, and they are. In fact, the dynamics of h_r for parameters in this interval are not fully understood. A bifurcation diagram for this interval is shown in Figure 10.8. Note the apparent period five and period three orbits. If we investigate a region around any of these orbits, we discover another bifurcation diagram apparently identical to the larger one we first examined in Figure 10.1. An enlargement of a region surrounding a period three point is shown in Figure 10.8. The reader is encouraged to use a computer to explore other regions of the bifurcation diagram. With a little effort, it should be possible to find a series of bifurcation diagrams, one contained within the other, all of which resemble Figure 10.1.

We complete our discussion of the period-doubling route to chaos by noting that the dynamics of h_r are already quite complicated by the time $r = 3.6786$. While Figure 10.8 makes it clear that the dynamics are complex by the time $r > 3.6786$, the sequence of period-doubling bifurcations has deteriorated into chaos long before then. In fact, the entire period-doubling cascade will have been completed by the time $r = 3.57$ and there will be a periodic point with prime period 2^n for each n whenever $r \geq 3.57$. Two techniques for demonstrating this are indicated in Exercises 10.4 and 10.8.

Readers who would like a more complete description of what happens to the dynamics of the logistic function as the parameter value increases from 3 to 4 are encouraged to consult Chapter 11 of *Fractals and Chaos* by Peitgen, Jürgens, and Saupe. The treatment found there is still heuristic but more extended than ours. Another treatment of this topic can be found in *An Introduction to Chaotic Dynamical Systems* by R. Devaney. Both texts are listed in the references.

Exercise Set 10

10.1 Let $h_r = rx(1-x)$ and $I_r = [\frac{1}{r}, \frac{r-1}{r}]$. Show that $h_r^2 : I_r \to I_r$ when $2 < r < 3.67$.

Hint: Refer to the discussion on page 98 for help.

10.2 Use a computer graphics package to demonstrate geometrically that there is another period-doubling bifurcation from period four to period eight inside the boxes shown in Figure 10.6. Explain why your graphs imply there is a such a bifurcation. At what parameter value does the bifurcation occur? Identify boxes on the graph of h^8 within which you would expect another period-doubling bifurcation.

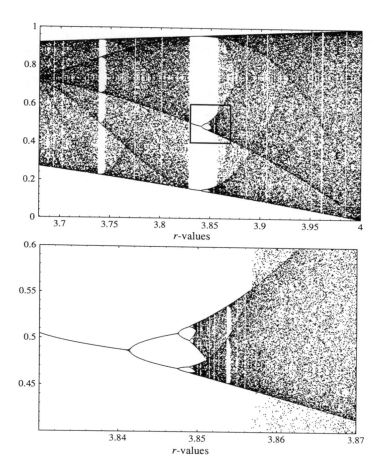

FIGURE 10.8. The top diagram is a bifurcation diagram for h_r in which r varies from 3.679 to 4. Note the vertical strips of white space. These strips correspond to attracting periodic points. Two of the most obvious occur at parameter values for which there are period five and period three attracting points. A box is drawn about a portion of the diagram in which there is a period three periodic point. The bottom diagram is an enlargement of this region. Note the similarity of this region to the bifurcation diagram shown in Figure 10.1. Given our discussion, the reader should be able to explain why this similarity is to be expected.

10.3 Use a computer graphics package to investigate the properties of h_r^3. Demonstrate graphically an attracting prime period three point of h_r. For which parameter values is there an attracting prime period three point? Find a box containing an attracting prime period three point in which the graph of h_r^3 is similar to the graph of h_s on the unit square for some s. Does a period-doubling bifurcation occur in this box? If so, what is the period of the new attracting fixed point? At what parameter value does it appear? Support your statements with graphs and explanations.

10.4 a) Draw a graph of h_r^2 for a variety of parameter values in the range $2 \leq r \leq 4$. How does the graph of h_r^2 change as r increases? Explain why your answer is correct. Use the results to argue that if h_r has a periodic point of prime period two when $r = r_0$, then it has a periodic point with prime period two whenever $r \geq r_0$.

b) Repeat part (a) for h_r^4.

c) Generalize the statements in part (a) to h_r^n. Explain why the result holds.

∗ d) Use part (c), Sarkovskii's Theorem, and a bifurcation diagram to demonstrate that the cascade of period-doubling bifurcations must have been completed before $r = 3.575$. That is, show there are orbits with prime period 2^n for all n whenever $r \geq 3.575$. Can you do the same for $r \geq 3.57$?

10.5 Let $h_r(x) = rx(1-x)$.

a) Show that $h_r(x)$ has an attracting orbit with prime period two when $3 < r < 1 + \sqrt{6}$.

Hint: Use the fact, which we demonstrated in Exercise 6.2, that if p_1 and p_2 are the prime period two points, then

$$(h^2)'(p_1) = (h^2)'(p_2) = h'(p_1) \cdot h'(p_2).$$

You will first need to solve the equation $h^2(x) = x$ to get expressions for p_1 and p_2 in terms of r. This equation has four roots, but we know two of them are the fixed points of $h(x)$ that occur at 0 and $\frac{r-1}{r}$. Thus, finding the roots reduces to solving a quadratic equation.

∗ b) Show that all of the points in $[0,1]$ that are not eventually fixed are in the stable set of one of the two prime period two points when $3 < r < 1 + \sqrt{6}$.

c) Show that a period-doubling bifurcation occurs at $r = 1 + \sqrt{6}$. In particular, show that the period two orbit is hyperbolic when $r = 1 + \sqrt{6}$ and that there is an attracting periodic point with prime period four when r is slightly larger than $1 + \sqrt{6}$. Through which parameter values will the attracting prime period four orbit persist?

Hint: It may help to review Exercise 7.3.

10.6 In Figure 10.4, we see that when $r = 3.8$, the point $\frac{1}{2}$ leaves the interval $[\frac{1}{r}, \frac{r-1}{r}]$ under iteration of $h^2(x)$ and then returns. Find a number in the interval $[\frac{r-1}{r}, \frac{r+\sqrt{r^2-4}}{2r}]$ that leaves the interval under iteration of $h^2(x)$ and then returns. Why is the existence of such a point significant?

Hint: It may help to review the discussion on pages 97–98.

•10.7 THE QUADRATIC MAP REVISITED:

Let $q_c(x) = x^2 + c$. Use a topological conjugacy to show that the only periodic points of q are a single attracting fixed point and a repelling fixed point when $-3/4 < c < 1/4$. However, when $c < -3/4$ we see a cascade of period-doubling bifurcations followed by chaos.

Hint: It may help to review Exercise 9.5.

10.8 FEIGENBAUM'S CONSTANT:

a) Let $q_c(x) = x^2 + c$ and define c_n to be the parameter value at which the period 2^n attracting periodic point of q_c splits into a period 2^{n+1} attracting periodic point. Estimate c_n for $n \leq 6$.

Hint: One way to do this is to use the fact that the critical point of $h(x)$ converges to the attracting periodic orbit under iteration of $h(x)$. By using care, we can determine at which parameter values period doubling occurs and new attracting periodic orbits appear.

b) Let $d_n = \dfrac{c_n - c_{n-1}}{c_{n+1} - c_n}$ and use the estimates of c_n found in part (a) to approximate d_n for $n \leq 5$. If you did the calculations carefully, you should find that the sequence of d_n's appears to converge. (Mitchell Feigenbaum showed that it converges to approximately 4.699 in 1978. This value is now called *Feigenbaum's constant*.)

Note: Another method of determining Feigenbaum's constant is outlined in Section 11.2 of *Chaos and Fractals* by Peitgen, Jürgens, and Saupe.

c) Use Feigenbaum's constant, the values c_n, and the fact that

$$\sum_{n=0}^{\infty} x^n = \frac{1}{1-x}$$ to estimate the value of r at which the period-doubling cascade will have been completed.

d) The amazing aspect of Feigenbaum's constant is that it is universal. By this we mean that if $S_c(x)$ is a parametrized family of functions which undergoes a period-doubling cascade in the same manner as the logistic function, c_n is the parameter value at which the bifurcation from period 2^n to period 2^{n+1} occurs, and $d_n = \frac{c_n - c_{n-1}}{c_{n+1} - c_n}$, then $\lim_{n \to \infty} d_n$ is equal to Feigenbaum's constant.

Show numerically that this is true for the family of functions $S_c(x) = c \sin x$ on the interval $[0, \pi]$. In creating the bifurcation diagram, assume that if $S_c(x)$ has an attracting periodic point, then a critical value of $S_c(x)$ must be in the stable set of one of its iterates. Recall that the critical value of a function is a point at which the derivative is zero.

10.9 Use a computer to find a sequence of ten bifurcation diagrams for h_r, one contained inside the other, each of which is similar to the diagram shown in Figure 10.1.

*10.10 What causes the shadow curves seen in the bifurcation diagrams of Figure 10.8?

Note: An interesting discussion of the shadow curves can be found in section 11.4 of the book *Fractals and Chaos* by Peitgen, Jürgens, and Saupe.

11

The Logistic Function Part IV: Symbolic Dynamics

We begin this chapter with a discussion of metric spaces and symbolic dynamics. As we proceed through the discussion, it may seem that the dynamics of the shift map on symbol space is an odd place to look for a deeper understanding of the logistic map, but we shall see that it is precisely the tool we need. In particular, we culminate the chapter by using symbolic dynamics to show that when $r > 2 + \sqrt{5}$, then there is a point in the set Λ whose orbit under iteration of $h_r(x) = rx(1-x)$ is dense in Λ.

11.1. Symbolic Dynamics and Metric Spaces

DEFINITION 11.1. *The set of all infinite sequences of 0's and 1's is called the sequence space of 0 and 1 or the symbol space of 0 and 1 and is denoted by* Σ_2. *More precisely,* $\Sigma_2 = \{(s_0 s_1 s_2 \ldots) \mid s_i = 0 \text{ or } s_i = 1 \text{ for all } i\}$. *We often refer to elements of* Σ_2 *as points in* Σ_2.

The concept of distance has proven to be useful when working on the real line, and we use it again in symbol space.

DEFINITION 11.2. *Let* $s = s_0 s_1 s_2 s_3 \ldots$ *and* $t = t_0 t_1 t_2 t_3 \ldots$ *be points in* Σ_2. *We denote the distance between* s *and* t *as* $d[s,t]$ *and define it by*

$$d[s,t] = \sum_{i=0}^{\infty} \frac{|s_i - t_i|}{2^i}.$$

Since $|s_i - t_i|$ is either 0 or 1,

$$0 \le d[s,t] \le \sum_{i=0}^{\infty} \frac{1}{2^i} = 2.$$

Hence, the furthest distance two points can be apart in Σ_2 is 2 and $d[s,t]$ is a number between 0 and 2. Also, it is clear that $d[s,t] = 0$ if and only if $s = t$. The distance function just described is an example of a metric. In general, we define a metric by

DEFINITION 11.3. *Let X be a set and let d be a function from the set of all ordered pairs of elements of X into the real numbers. If the following conditions hold for all x, y, and z in X, then d is a metric on X.*

   a) $d[x,y] \ge 0$ *and* $d[x,y] = 0$ *if and only if* $x = y$.
   b) $d[x,y] = d[y,x]$.
   c) $d[x,y] \le d[x,z] + d[z,y]$.

We are already familiar with one metric: the distance between two points in the real numbers is the absolute value of their difference. For example, the distance between the numbers 2 and 3 is $|2-3| = 1$, and the distance between 2 and -5 is $|2-(-5)| = 7$. It should be clear that this distance function satisfies conditions (a) and (b) of the definition. The reader may recognize condition (c) in Definition 11.3 as the Triangle Inequality, which was proven for the metric on the real numbers in Exercise 2.8. In the exercises at the end of this chapter, we show that the distance function in symbol space is a metric and evaluate several other functions as possible metrics.

The obvious question now is "Why do we care?". If we refer back to Chapter 3, we see that all of our topological definitions and the definition of continuity are based on the idea of distance. So once we have defined a metric on a set, we can define a topology for the set. That is, we can decide which subsets are open, which are closed, what a convergent sequence is in the set, and which functions of the set are continuous. All of these definitions are brought together here.

DEFINITION 11.4. *Let X be a set and d be a metric on the set.*

   a) *A subset U of X is open if for each x in U there exists $\epsilon > 0$ such that $d[x,y] < \epsilon$ implies y is in U.*
   b) *Let $\epsilon > 0$ and x be in X. The set $N_\epsilon(x) = \{y \text{ in } X \mid d[x,y] < \epsilon\}$ is called a neighborhood of x.*
   c) *Let $x_1, x_2, x_3, \ldots$ be a sequence of elements of X. The sequence converges to x in X if for each $\epsilon > 0$, there exists an integer N such that if $k \ge N$, then $d[x, x_k] < \epsilon$.*

d) *Let S be a subset of X. Then the point x in X is an accumulation point (or a limit point) of S, if every neighborhood of x contains an element of S that is distinct from x.*

e) *A subset of X is closed if it contains all of its accumulation points.*

f) *Let B be a subset of X and A be a subset of B. Then A is dense in B if every point of B is an accumulation point of A, a point of A, or both. In other words, if A is dense in B and x is in B, then every neighborhood of x contains an element of A.*

g) *If Y is a set and d_2 is a metric on Y, then the function $f : X \to Y$ is continuous at the point x_0 in X if for every $\epsilon > 0$ there exists $\delta > 0$ such that*

if x is in X and $d[x_0, x] < \delta$, then $d_2[f(x_0), f(x)] < \epsilon$.

Note that $f(x_0)$ and $f(x)$ are elements of Y, so we must use the metric d_2 to measure the distance between them. A function is continuous if it is continuous at each point of its domain.

Notice the importance of distance in the previous definitions. The set U is open if the points near each point (within the distance ϵ of each point) are in U. The sequence $\{x_n\}$ converges to x if the point x_n is close to x (within the distance ϵ of x) when n is sufficiently large. The point x is an accumulation point of S if we can find a point of S as close to x as we want. A is dense in B if there is a point in A close (within the distance ϵ) to each point of B. Finally, the function f is continuous at x_0 if we can guarantee $f(x)$ is close to $f(x_0)$ whenever x is close enough to x_0.

All of the topological facts that we have shown to be true for the real numbers are also true for more general metric spaces. We gather together some of these facts in the following propositions. The proofs of all of these propositions follow directly from the definitions and are left as exercises. Students who wish to master the concepts of metric spaces and their topologies are urged to complete the proofs of all the propositions and study the examples that follow.

PROPOSITION 11.5. *The set U is open if and only if for each x in U there is a neighborhood of x that is completely contained in U.*

PROPOSITION 11.6. *Every neighborhood of a point in a metric space is an open set. That is, if X is a metric space, x is in X, and $\epsilon > 0$, then the neighborhood $N_\epsilon(x)$ is open.*

PROPOSITION 11.7. *Let X be a metric space with metric d. If x is in X and S is a subset of X, then the following statements are equivalent:*

a) *x is an accumulation point of S.*

b) *For each $\epsilon > 0$ there exists y in S such that $0 < d[x, y] < \epsilon$.*

   c) *If U is an open set containing x, then $U \cap S$ contains at least one point other than x.*

   d) *There is a sequence of points different from x and contained in S that converges to x.*

PROPOSITION 11.8. *The complement of an open set is closed. Conversely, the complement of a closed set is open.*

PROPOSITION 11.9. *Let X be a metric space with metric d, B be a subset of X, and A be a subset of B. Then the following statements are equivalent:*

   a) *A is dense in B.*

   b) *For all points b in B and all $\epsilon > 0$, there exists a in A such that $d[b, a] < \epsilon$.*

   c) *For each b in B, there is a sequence of points $a_1, a_2, a_3, \ldots$ contained in A that converges to b.*

   d) *Every open set in X that has a nonempty intersection with B contains an element of A.*

PROPOSITION 11.10. *Let X and Y be metric spaces and $f : X \to Y$ be a continuous function. If U is an open set in Y, then $f^{-1}(U)$ is an open set in X.*

Before moving on to a few examples, we pause to prove a useful lemma about Σ_2.

LEMMA 11.11. *Let s and t be elements of Σ_2. If the first $n+1$ digits in s and t are identical, then $d[s, t] \leq \frac{1}{2^n}$. On the other hand, if $d[s, t] \leq \frac{1}{2^n}$, then the first n digits in s and t are identical.*

PROOF. Let $s = s_0 s_1 s_2 \ldots$ and $t = t_0 t_1 t_2 \ldots$ be sequences in Σ_2. We note first that the first $n+1$ digits of s are $s_0, s_1, \ldots, s_n$. So, s and t agree on the first $n+1$ digits if and only if $s_i = t_i$ for $i \leq n$.

Now suppose $s_i = t_i$ for all $i \leq n$. Then

$$d[s, t] = \sum_{i=0}^{\infty} \frac{|s_i - t_i|}{2^i} = \sum_{i=0}^{n} \frac{0}{2^i} + \sum_{i=n+1}^{\infty} \frac{|s_i - t_i|}{2^i}$$

$$= \frac{1}{2^{n+1}} \sum_{i=0}^{\infty} \frac{|s_{i+n+1} - t_{i+n+1}|}{2^i} \leq \frac{1}{2^{n+1}} \sum_{i=0}^{\infty} \frac{1}{2^i} = \frac{1}{2^n}.$$

On the other hand, if there is $j < n$ such that $s_j \neq t_j$, then

$$d[s,t] = \sum_{i=0}^{\infty} \frac{|s_i - t_i|}{2^i} \geq \frac{1}{2^j} > \frac{1}{2^n}.$$

Consequently, if $d[s,t] \leq \frac{1}{2^n}$, then $s_j = t_j$ for all $j < n$, and the first n digits in s and t are identical. □

EXAMPLE 11.12.
a) The distance between $s = 000000\ldots$ and $t = 010101010\ldots$ is

$$\sum_{i=0}^{\infty} \frac{|s_i - t_i|}{2^i} = \frac{0}{2^0} + \frac{1}{2^1} + \frac{0}{2^2} + \frac{1}{2^3} + \frac{0}{2^4} + \frac{1}{2^5} + \cdots$$

$$= \frac{1}{2} \sum_{i=0}^{\infty} \frac{1}{4^i}$$

$$= \frac{1}{2} \left(\frac{1}{1 - \frac{1}{4}} \right) = \frac{2}{3}.$$

b) If S is the set of all elements of Σ_2 beginning with the sequence 011, then S is closed.

To demonstrate this, we show that all of the accumulation points of S are contained in S. If s is an accumulation point of S, then every neighborhood of s contains an element of S other than s. Consequently, there is s^* in $N_{\frac{1}{2^3}}(s) \cap S$ satisfying $s^* \neq s$. But then the definition of neighborhood implies $d[s^*, s] < \frac{1}{2^3}$ and Lemma 11.11 implies that the first three digits of s and s^* must agree. Since s^* is in S, the first three digits of s^* are 011. Hence, the first three digits of s are 011 and s is in S.

c) The set of all sequences in Σ_2 ending with an infinite string of 0's is dense in symbol space. That is, if

$$A = \{s_0 s_1 s_2 \ldots \text{ in } \Sigma_2 \mid \text{there is } N \text{ satisfying } s_i = 0 \text{ for all } i \geq N\},$$

then A is dense in Σ_2.

To show this, let t be in Σ_2 and $N_\epsilon(t)$ be a neighborhood of t. We must prove that there is s in $A \cap N_\epsilon(t)$. That is, we must show that there is s in A such that $d[s,t] < \epsilon$. Choose n so that $\frac{1}{2^n} < \epsilon$. Let s be the element of Σ_2 whose first $n+1$ digits are the same as t's first $n+1$ digits and whose remaining digits are all 0's. Then s is in A by definition and Lemma 11.11 implies $d[s,t] \leq \frac{1}{2^n} < \epsilon$. Hence, $N_\epsilon(t)$ contains an element of A, and we're done. □

114   11. The Logistic Function Part IV: Symbolic Dynamics

Now that we have a basic understanding of Σ_2, we are ready to examine the dynamics of the shift map on Σ_2.

DEFINITION 11.13. *The shift map $\sigma : \Sigma_2 \to \Sigma_2$ is defined by*

$$\sigma(s_0 s_1 s_2 \ldots) = s_1 s_2 s_3 \ldots.$$

In other words, the shift map "forgets" the first digit of the sequence. For example, $\sigma(01110101\ldots) = 1110101\ldots.$

There are several immediate observations we can make about the shift map.

PROPOSITION 11.14. *The shift map is continuous.*

PROOF. Let s be an element of Σ_2 and $\epsilon > 0$. We must show that there is $\delta > 0$ such that whenever $d[s,t] < \delta$, then $d[\sigma(s), \sigma(t)] < \epsilon$. Choose n so that $\frac{1}{2^n} < \epsilon$ and let $\delta = \frac{1}{2^{n+2}}$. If $d[s,t] < \delta$, then we know from Lemma 11.11 that s and t agree on the first $n+2$ digits. So $\sigma(s)$ and $\sigma(t)$ agree on the first $n+1$ digits and $d[\sigma(s), \sigma(t)] \leq \frac{1}{2^n} < \epsilon$ as desired. □

It should be clear by now that two points are close together if and only if their initial digits agree. The more digits they agree on before they differ, the closer together they are. Of course, this is exactly what Lemma 11.11 claims. Example 11.12 and the proof of Proposition 11.14 illustrate the use of this idea. We will use this idea again in the proof of part (a) of Proposition 11.15, and the reader is given an opportunity to use it in proving parts (c) and (d).

PROPOSITION 11.15. *The shift map has the following properties:*
   a) *The set of periodic points of the shift map is dense in Σ_2.*
   b) *The shift map has 2^n periodic points of period n.*
   c) *The set of eventually periodic points of the shift map that are not periodic is dense in Σ_2.*
   d) *There is an element of Σ_2 whose orbit is dense in Σ_2. That is, there is s^* in Σ_2 such that the set $\{s^*, \sigma(s^*), \sigma^2(s^*), \sigma^3(s^*), \ldots\}$ is dense in Σ_2.*
   e) *The set of points that are neither periodic nor eventually periodic is dense in Σ_2.*

PROOF. a) Suppose $s = s_0 s_1 s_2 \ldots$ is a periodic point of σ with period k. Then $\sigma^n(\sigma^k(s)) = \sigma^n(s)$. Since $\sigma^n(s)$ "forgets" the first n digits of s, we see that

$$\sigma^n(\sigma^k(s_0 s_1 s_2 \ldots)) = s_{n+k} s_{n+k+1} s_{n+k+2} \ldots$$
$$= s_n s_{n+1} s_{n+2} \ldots = \sigma^n(s_0 s_1 s_2 \ldots)$$

and $s_{n+k} = s_n$ for all n. This implies that s is a periodic point with period k if and only if s is a sequence formed by repeating the k digits $s_0 s_1 \ldots s_{k-1}$ infinitely often.

To prove that the periodic points of σ are dense in Σ_2, we must show that for all points t in Σ_2 and all $\epsilon > 0$, there is a periodic point of σ contained in $N_\epsilon(t)$. From the preceding discussion, we see that this means we need to find a sequence in $N_\epsilon(t)$ that is formed by repeating the initial k digits of the sequence infinitely often. But, if $t = t_0 t_1 t_2 t_3 \ldots$ and we choose n so that $\frac{1}{2^n} < \epsilon$, then we can let $s = t_0 t_1 \ldots t_n t_0 t_1 \ldots t_n t_0 t_1 \ldots$. As t and s agree on the first $n+1$ digits, Lemma 11.11 implies that $d[s,t] \leq \frac{1}{2^n} < \epsilon$. Thus, s is in $N_\epsilon(t)$ and s is a periodic point by construction.

The proofs of parts (b) and (c) are left as exercises.

d) The sequence which begins with 0 1 00 01 10 11 and then includes all possible blocks of 0 and 1 with three digits (there are eight), followed by all possible blocks of 0 and 1 with four digits (there are sixteen), and so forth is called the *Morse sequence*. We leave as an exercise the proof that the orbit of this sequence is dense in Σ_2.

e) Since the set of nonperiodic points includes as a subset the orbit of the Morse sequence, the truth of part (e) of this proposition follows from part (d). □

Proposition 11.15 provides us with the tools we need to characterize the behavior of the shift map σ on Σ_2. We show in Proposition 11.17 that parts (a) and (d) of Proposition 11.15 imply that σ is topologically transitive on Σ_2. The definition of topological transitivity for functions on metric spaces is analogous to the definition for functions of the real numbers, which we introduced in Chapter 8.

DEFINITION 11.16. *Let D be a subset of a metric space. The function $f : D \to D$ is topologically transitive on D if for any open sets U and V that intersect D there is z in $U \cap D$ and a natural number n such that $f^n(z)$ is in V. Equivalently, f is topologically transitive on D if for any two points x and y in D and any $\epsilon > 0$, there is z in D such that $d[z, x] < \epsilon$ and $d[f^n(z), y] < \epsilon$ for some n.*

PROPOSITION 11.17. *Let D be a subset of a metric space and $f : D \to D$. If the periodic points of f are dense in D and there is a point whose orbit under iteration of f is dense in D, then f is topologically transitive on D.*

PROOF. Let $f : D \to D$. Assume that the periodic points of f are dense in D and that there is a point in D whose orbit is dense in D. That is, there exists x_0 in D such that every open set that has a nonempty intersection with D contains an element of the form $f^k(x_0)$.

We begin our proof by making two observations. First, if D is a finite set, then D must consist of a single periodic orbit. Thus, when D is a finite set, f must be topologically transitive on D. Second, if D is an infinite set, then no iterate of any point with a dense orbit can be periodic. That is, if x_0 is either periodic or eventually periodic, then x_0 cannot have a dense orbit. This follows from the fact that a finite subset of an infinite set cannot be dense in the set. The reader is asked to create a detailed proof of these two observations in Exercise 11.16.

Now we wish to show that f is topologically transitive on D. Let U and V be open sets that have nonempty intersections with D. We must show that there is x in $U \cap D$ and a natural number n such that $f^n(x)$ is in V. Given the first observation in the preceding paragraph we need only consider the case where D is infinite. By hypothesis, there is a point in D whose orbit is dense in D. Let x_0 be such a point. Choose k such that $f^k(x_0)$ is in U. To complete the proof, it suffices to show that there is a natural number m such that $m > k$ and $f^m(x_0)$ is in V. In this case, we can let $x = f^k(x_0)$ and $n = m - k$. Then x is in U and the point

$$f^n(x) = f^{m-k}(x) = f^{m-k}(f^k(x_0)) = f^m(x_0)$$

is in V. To show that such an m exists, it suffices to prove that there are infinitely many iterates of x_0 in V. In this case, since there are only finitely many natural numbers less than k, there are infinitely many iterates of the form $f^m(x_0)$ where $m > k$. We use contradiction to show that there are infinitely many iterates of x_0 in V.

Suppose that V contains only finitely many iterates of x_0 and that p is a periodic point in V. We know that p exists since the periodic points of f are dense in D. Our second observation implies that p is not in the orbit of x_0. So if $f^j(x_0)$ is one of the iterates of x_0 contained in V, then the distance $d[f^j(x_0), p]$ is positive. Since there are only finitely many iterates of x_0 contained in V, there is a smallest such distance, which we will call ϵ. Then the neighborhood $N_\epsilon(p)$ is open and has nonempty intersection with D. Also, since V and $N_\epsilon(p)$ are open, the intersection $V \cap N_\epsilon(p)$ is open and contains p, so its intersection with D is nonempty. (The reader is asked to prove that the intersection of open sets is open in Exercise 11.13.) Since the orbit of x_0 is dense in D, there must exist i such that $f^i(x_0)$ is in $V \cap N_\epsilon(p)$. But this implies that $f^i(x_0)$ is in V, and so by the definition of ϵ, we know that $d[f^i(x_0), p] \geq \epsilon$. Of course, this means that $f^i(x_0)$ is not in $N_\epsilon(p)$ and so can't be in $V \cap N_\epsilon(p)$, a contradiction. Since we arrived at the contradiction by assuming that V contains only finitely many iterates of x_0, we conclude that V contains infinitely many iterates of x_0, and the proof is complete. □

In Chapter 8, we showed that if $f : D \to D$ is continuous, D is an infinite subset of the real numbers, the periodic points of f are dense in D, and f is topologically transitive on D, then f is chaotic on D. (See Theorem 8.16.) Before presenting the analogous theorem for metric spaces, we formalize the relevant definitions.

DEFINITION 11.18. *Let D be a subset of a metric space with metric d. The function $f : D \to D$ exhibits sensitive dependence on initial conditions if there exists a $\delta > 0$ such that for any x in D and any $\epsilon > 0$, there is a y in D and a natural number n such that $d[x, y] < \epsilon$ and $d[f^n(x), f^n(y)] > \delta$.*

DEFINITION 11.19. *Let D be a subset of a metric space. The function $f : D \to D$ is chaotic if*

   a) *the periodic points of f are dense in D,*

   b) *f is topologically transitive, and*

   c) *f exhibits sensitive dependence on initial conditions.*

Note that the definitions of dynamical concepts for metric spaces are completely analogous to those of functions of the real numbers, the only difference being the introduction of the metric, $d[x, y]$, in place of the standard distance, $|x - y|$, on the real numbers. Consequently, it is no surprise that many of the theorems for functions of the real numbers have analogs for more general functions defined on metric spaces. In particular, we note that when the domain of a continuous function is infinite, then the density of periodic points and topological transitivity imply sensitive dependence on initial conditions.

THEOREM 11.20. *Let D be an infinite subset of a metric space and $f : D \to D$ be continuous. If f is topologically transitive on D and the periodic points of f are dense in D, then f is chaotic on D.*

The proof of Theorem 11.20 is analogous to the proof of Theorem 8.16. In fact, since $|a - b|$ represents the distance from a to b in the real numbers, to create a proof of Theorem 11.20 it suffices to replace references to expressions of the form $|a - b|$ by expressions of the form $d[a, b]$ in the proof of Theorem 8.16. The reader may wish to review some of the other theorems and propositions about the dynamics of functions of the real numbers to determine which have analogs for more general functions.

While Proposition 11.15, Proposition 11.17, and Theorem 11.20 together imply that the shift map exhibits sensitive dependence on initial conditions, we can easily show that it exhibits sensitive dependence on initial conditions directly from the definition. In fact, we show in the following proposition that in every neighborhood of a point there is another point whose iterates are eventually a distance 2 away from the iterates of the first point. Thus, after a finite number of iterations, the points are as far apart as two sequences can be in Σ_2, even though we can initially choose them to be as close together as we want. This is an extreme example of sensitive dependence on initial conditions.

PROPOSITION 11.21. *Let s be any point in Σ_2 and $\epsilon > 0$. Then there is t in Σ_2 and N such that $d[s,t] < \epsilon$ and $d[f^n(s), f^n(t)] = 2$ whenever n is larger than N.*

PROOF. Let $s = s_0 s_1 s_2 \ldots$ be an element of Σ_2 and $\epsilon > 0$. We will show that there is t in Σ_2 and a natural number N such that $d[s,t] < \epsilon$, but $d[\sigma^n(t), \sigma^n(s)] = 2$ whenever $n \geq N$. We begin by choosing N so that $\frac{1}{2^N} < \epsilon$. We then choose $t = t_0 t_1 t_2 \ldots$ so that the first $N+1$ digits of t are the same as the first $N+1$ digits of s and the other digits of t are all different from the corresponding digits of s. That is, $s_i = t_i$ if and only if $i \leq N$. With s and t so defined, we know from Lemma 11.11 that $d[s,t] \leq \frac{1}{2^N} < \epsilon$. If $n > N$, then $\sigma^n(s)$ and $\sigma^n(t)$ differ at each digit since $\sigma^n(s) = s_n s_{n+1} s_{n+2} \ldots$ and $\sigma^n(t) = t_n t_{n+1} t_{n+2} \ldots$. Thus,

$$d[\sigma^n(t), \sigma^n(s)] = \sum_{i=0}^{\infty} \frac{|s_{i+n} - t_{i+n}|}{2^i} = \sum_{i=0}^{\infty} \frac{1}{2^i} = 2.$$

Therefore, we have found t and N so that $d[s,t] < \epsilon$ and $d[\sigma^n(s), \sigma^n(t)] = 2$ whenever n is larger than N. □

We close this section by noting, as the reader undoubtedly already has, that the shift map is chaotic on Σ_2.

COROLLARY 11.22. *The shift map σ is chaotic on Σ_2.*

PROOF. From part (a) of Proposition 11.15, we know that periodic points of σ are dense in Σ_2. Proposition 11.17 and parts (a) and (d) of Proposition 11.15 guarantee that σ is topologically transitive on Σ_2. Proposition 11.21 implies that $\sigma : \Sigma_2 \to \Sigma_2$ exhibits sensitive dependence on initial conditions. Thus, σ is chaotic on Σ_2. □

11.2. Symbolic Dynamics and the Logistic Function

Let $h_r(x) = rx(1-x)$ and $\Lambda = \{x \mid h^n(x) \text{ is in } [0,1] \text{ for all } n\}$. We complete this chapter by demonstrating that the map $\sigma : \Sigma_2 \to \Sigma_2$ and the function $h_r : \Lambda \to \Lambda$ are topologically conjugate when $r > 2 + \sqrt{5}$. Once this has been done, we show that when $r > 2 + \sqrt{5}$ there is a point in Λ whose orbit under iteration of h_r is dense in Λ, an interesting and counterintuitive result. As with many other results we have shown for the logistic function, it is true that such a point with a dense orbit in Λ exists when $4 < r \leq 2 + \sqrt{5}$, but again the proof is more difficult. The interested reader is referred to Section 3.5 of the text *Dynamical Systems: Stability, Symbolic Dynamics, and Chaos* by C. Robinson, where it is demonstrated that $\sigma : \Sigma_2 \to \Sigma_2$ and $h_r : \Lambda \to \Lambda$ are topologically conjugate whenever $r > 4$.

We note first that topological conjugacies for metric spaces are defined as one would expect. In fact, looking back at Definition 9.1, we see that the definition given there applies to functions of metric spaces as well as functions of the real numbers. That is, a topological conjugacy between two functions defined on metric spaces is a homeomorphism of the domains of the function that commutes with the functions.

We next define a function $\psi : \Lambda \to \Sigma_2$, which we shall demonstrate is a topological conjugacy. Set $I_0 = [0, \frac{1}{2} - \frac{\sqrt{r^2-4r}}{2r}]$ and $I_1 = [\frac{1}{2} + \frac{\sqrt{r^2-4r}}{2r}, 1]$. Recall from Proposition 8.1 that

$$\Lambda_1 = \{x \mid h(x) \text{ is in } [0,1]\}$$
$$= I_0 \cup I_1.$$

Since Λ is a subset of Λ_1, Λ is contained in $I_0 \cup I_1$. For each x in Λ, define the sequence $\psi(x) = s_0 s_1 s_2 \ldots$ in Σ_2 so that for each n, $s_n = 0$ if and olny if $h^n(x)$ is in I_0 and $s_n = 1$ if and olny if $h^n(x)$ is in I_1. For example, if $r = 5$, then $h^0(\frac{1-\sqrt{5}}{2}) = \frac{1-\sqrt{5}}{2}$ is in I_0, $h^1(\frac{1-\sqrt{5}}{2}) = 1$ is in I_1, and $h^k(\frac{1-\sqrt{5}}{2}) = 0$ is in I_0 for all $k \geq 2$. So $\psi(\frac{1-\sqrt{5}}{2}) = 010000\ldots$. We can think of $\psi(x)$ as the itinerary of x. Clearly, ψ is well-defined. It remains to show that ψ is a topological conjugacy.

THEOREM 11.23. *The function $\psi : \Lambda \to \Sigma_2$ is a topological conjugacy. That is,*

a) *ψ is one-to-one and onto,*

b) *ψ is continuous,*

c) *ψ^{-1} is continuous, and*

d) *$\psi \circ h = \sigma \circ \psi$.*

PROOF. We begin our proof by defining a sequence of intervals,

$$I_{s_0} \supset I_{s_0 s_1} \supset I_{s_0 s_1 s_2} \supset \cdots \supset I_{s_0 s_1 s_2 s_3 \ldots s_n} \supset \cdots,$$

for each element of $s_0 s_1 s_2 \ldots$ in Σ_2. The intervals are chosen so that $I_{s_0 s_1 s_2 s_3 \ldots s_n}$ is a subset of Λ_n for each n. Once this is done, we will prove each of the parts of Theorem 11.23 in turn.

Let h, I_0, and I_1 be defined as in the discussion preceding the statement of the theorem and let $s_0 s_2 s_2 \ldots$ be an element of Σ_2. We define the set

$I_{s_0s_1s_2...s_n}$ for each n by induction on n. Let $I_{s_0} = I_0$ if $s_0 = 0$ and $I_{s_0} = I_1$ if $s_0 = 1$. Now suppose that $I_{s_0s_1s_2...s_{n-1}}$ is defined. Then

$$I_{s_0s_1s_2...s_n} = \{x \text{ in } I_{s_0s_1s_2...s_{n-1}} \mid h^n(x) \text{ is in } I_{s_n}\},$$

where $I_{s_n} = I_0$ if $s_n = 0$ and $I_{s_n} = I_1$ if $s_n = 1$. In other words, $I_{s_0s_1s_2...s_n}$ is the set of points in Λ that satisfy the condition that $h^k(x)$ is in I_{s_k} for each k less than or equal to n. Again, $I_{s_k} = I_0$ if $s_k = 0$ and $I_{s_k} = I_1$ if $s_k = 1$. We see that the points in $I_{s_0s_1s_2...s_n}$ have a common itinerary for the first n iterates of h. That is, if x and y are in $I_{s_0s_1s_2...s_n}$, then for all $k \leq n$ we know that $h^k(x)$ is in I_0 if and only if $h^k(y)$ is in I_0. Similarly, $h^k(x)$ is in I_1 if and only if $h^k(y)$ is in I_1. We claim that $I_{s_0s_1s_2...s_n}$ is one of the intervals comprising Λ_{n+1} where

$$\Lambda_{n+1} = \{x \mid h^{n+1}(x) \text{ is in } [0,1]\}.$$

We note that this definition of Λ_{n+1} is the one used in Chapter 8. (See equation 8.1 on page 70.) We prove the claim by induction on n.

The set I_{s_0} is either I_0 or I_1 by definition. By Proposition 8.1, Λ_1 is comprised of I_0 and I_1, so our claim is true for the case $n = 0$. Now suppose $I_{s_0s_1...s_{n-1}}$ is one of the intervals comprising Λ_n. We will show this implies that $I_{s_0s_1...s_n}$ is one of the intervals in Λ_{n+1}. Note first that $I_{s_0s_1...s_n}$ is a subset of $I_{s_0s_1...s_{n-1}}$ so it suffices to determine which portions of $I_{s_0s_1...s_{n-1}}$ belong to $I_{s_0s_1...s_n}$.

Let $I_{s_0s_1...s_{n-1}} = [a,b]$. Proposition 8.1 states that $h^n([a,b]) = [0,1]$ and h^n is monotone on $[a,b]$. We assume that h^n is increasing on $[a,b]$; the proof is similar when h^n is decreasing. By the Intermediate Value Theorem, there exist c_1 and c_2 such that $a < c_1 < c_2 < b$, $h^n(c_1) = \frac{1}{2} - \frac{\sqrt{r^2-4r}}{2r}$, and $h^n(c_2) = \frac{1}{2} + \frac{\sqrt{r^2-4r}}{2r}$. Consequently,

$$h^n([a,c_1]) = \left[0, \frac{1}{2} - \frac{\sqrt{r^2-4r}}{2r}\right] = I_0,$$

$$h^n((c_1,c_2)) = \left(\frac{1}{2} - \frac{\sqrt{r^2-4r}}{2r}, \frac{1}{2} + \frac{\sqrt{r^2-4r}}{2r}\right),$$

and

$$h^n([c_2,b]) = \left[\frac{1}{2} + \frac{\sqrt{r^2-4r}}{2r}, 1\right] = I_1.$$

So, if $s_n = 0$, then $I_{s_0s_1...s_n} = [a,c_1]$ and if $s_n = 1$, then $I_{s_0s_1...s_n} = [c_2,b]$. As this is the same way we derived the intervals of Λ_{n+1} in the proof of Proposition 8.1, we see that $I_{s_0s_1...s_n}$ is one of the intervals comprising Λ_{n+1}.

a) Let $s = s_0 s_1 s_2 s_3 \ldots$ be an element of Σ_2. To show that ψ is one-to-one and onto we must show that $\psi^{-1}(s)$ contains exactly one point. But if x is in $\psi^{-1}(s)$, then x is in $I_{s_0 s_1 \ldots s_n}$ for all n. Thus,

$$\psi^{-1}(s) = \bigcap_{n=0}^{\infty} I_{s_0 s_1 \ldots s_n}.$$

We need to show that this intersection is not empty and contains exactly one point. Let $I_{s_0 s_1 \ldots s_n} = [a_n, b_n]$. Suppose that $\psi^{-1}(s)$ contains two points x and y. Then $|x - y| \leq |b_n - a_n|$ for all n since x and y are in each of the intervals $[a_n, b_n]$. However, Lemma 8.6 implies that $|b_n - a_n|$ approaches 0 as n goes to infinity. So it must be that $|x - y| = 0$ and $x = y$. Consequently, we see that $\psi^{-1}(s)$ contains no more than one point.

It remains to show that $\psi^{-1}(s)$ contains at least one point. This follows immediately from the well-known Nested Interval Theorem of mathematical analysis, which states that the intersection of a nested set of closed intervals is not empty. (See theorem 2.38 of W. Rudin's *Principles of Mathematical Analysis*, third edition.) Recall that the intervals $I_{s_0 s_1 \ldots s_n}$ are denoted by $[a_n, b_n]$. The proof of the Nested Interval Theorem depends on the principle that there exists a unique real number larger than or equal to every a_n and smaller than or equal to any other number that is also greater than or equal to all the a_n. Such a number is called the *least upper bound* of the set $\{a_1, a_2, \ldots, a_n, \ldots\}$. Now, since $a_n \leq a_m \leq b_m$ whenever $m \geq n$ and $a_n \leq b_n \leq b_m$ whenever $m \leq n$, we see that each b_m is greater than every a_n. So the least upper bound, a, is less than or equal to every b_m. Therefore, $a_n \leq a \leq b_n$ for each n and a is in every $[a_n, b_n] = I_{s_0 s_1 \ldots s_n}$. It follows that a is in $\psi^{-1}(s)$ and we have completed the proof of part (a).

b) Let $\epsilon > 0$ and x be in Λ. To show that ψ is continuous at x, we must find $\delta > 0$ so that if $|x - y| < \delta$, then $d[\psi(x), \psi(y)] < \epsilon$. Choose n so that $\frac{1}{2^n} < \epsilon$. Then Lemma 11.11 implies that it is sufficient to show there exists a $\delta > 0$ such that the sequences $\psi(x)$ and $\psi(y)$ agree on the first $n+1$ digits whenever $|x - y| < \delta$. In other words, if $\psi(x) = s_0 s_1 s_2 \ldots$, then x is in $I_{s_0 s_1 s_2 \ldots s_n}$ and we need to show that y is in $I_{s_0 s_1 s_2 \ldots s_n}$ whenever y is in Λ and $|x - y| < \delta$.

To do this, let $[a_1, b_1], [a_2, b_2], \ldots, [a_{2^{n+1}}, b_{2^{n+1}}]$ be the 2^{n+1} intervals in Λ_{n+1} indexed so that $b_{i-1} < a_i$ for all i. Set δ equal to the smallest distance between consecutive intervals. That is, δ is the smallest number of the form $|a_i - b_{i-1}|$ where $1 < i \leq 2^{n+1}$. Since the intervals are disjoint and finite in number, δ is positive. Also, if x and y are in Λ_{n+1} and $|x - y| < \delta$, then it follows that x and y are in the same interval of Λ_{n+1}. As $I_{s_0 s_1 s_2 \ldots s_n}$ is an interval in Λ_{n+1} and x is in $I_{s_0 s_1 s_2 \ldots s_n}$, we see that y must be in $I_{s_0 s_1 s_2 \ldots s_n}$, and the proof of part (b) is complete.

c) The proof that ψ^{-1} is continuous is left as an exercise.

d) It remains to show that $\psi \circ h = \sigma \circ \psi$. Let x be an element of Λ and $\psi(x) = s_0 s_1 s_2 \ldots$. As we demonstrated in the proof of part (a), x is the unique point in the intersection

$$\bigcap_{n=0}^{\infty} I_{s_0 s_1 s_2 \ldots s_n} = \bigcap_{n=0}^{\infty} \{x \mid h^k(x) \text{ is in } I_{s_k} \text{ for all } k \leq n\}.$$

It follows that $h(x)$ is the unique point in the intersection

$$I_{s_1 s_2 s_3 \ldots} = \bigcap_{n=1}^{\infty} I_{s_1 s_2 \ldots s_n} = \bigcap_{n=0}^{\infty} \{x \mid h^{k+1}(x) \text{ is in } I_{s_k} \text{ for all } k \leq n\}.$$

So $\psi(h(x)) = s_1 s_2 s_3 \cdots = \sigma(\psi(x))$, and we are done. □

To use Theorem 11.23, we need to know that topological conjugacies between functions of metric spaces preserve the dynamical properties of these functions. That is, we need to know that Theorem 9.3 holds for functions defined on subsets of a metric space. This is indeed the case, and we will not strain the readers patience by revisiting each detail of the theorem. Instead, we note that the proof of the more general result is identical to that of Theorem 9.3 with the obvious changes made to allow for the metric. The reader is invited to reread Theorem 9.3 and its proof and note the places where changes need to be made; there are very few.

COROLLARY 11.24. *Let* $r > 2 + \sqrt{5}$, $h(x) = rx(1-x)$, *and*

$$\Lambda = \{x \mid h^n(x) \text{ is in } [0,1] \text{ for all } n\}.$$

Then Λ contains a point whose orbit is dense in Λ.

PROOF. The result follows immediately from Theorem 11.23, part (d) of Proposition 11.15, and the metric space analog of Theorem 9.3. □

Exercise Set 11

11.1 a) Show that the distance function defined on Σ_2 in Definition 9.2 is a metric.

b) Let $z = 00000000\ldots$, $r = 11111\ldots$, and $s = 001001001001\ldots$. Calculate $d[z, r]$, $d[r, s]$, and $d[z, s]$.

c) Show that the largest distance two points can be apart in Σ_2 is 2.

Exercise Set 11   123

11.2 Let C be the set of all ordered pairs of real numbers (i.e., C is the Cartesian plane).

- a) Show that $d_1[(x_1, y_1), (x_2, y_2)] = \sqrt{(x_1 - x_2)^2 + (y_1 - y_2)^2}$ is a metric.
  Let $\mathbf{y} = (1, 2)$ be a point in C and graph the set
  $$N_1(\mathbf{y}) = \{\mathbf{x} \text{ in } C \mid d_1[\mathbf{x}, \mathbf{y}] < 1\}.$$

b) Show that $d_2[(x_1, y_1), (x_2, y_2)] = |x_1 - x_2| + |y_1 - y_2|$ is a metric. Let $\mathbf{y} = (1, 2)$ be a point in C and graph the set
$$N_1(\mathbf{y}) = \{\mathbf{x} \text{ in } C \mid d_2[\mathbf{x}, \mathbf{y}] < 1\}.$$

11.3 Is the function $d_3[x, y] = (x - y)^2$ a metric on the real numbers?
Is the function $d_4[x, y] = \frac{|x-y|}{1+|x-y|}$ a metric on the real numbers?

11.4 Prove Propositions 11.5, 11.6, 11.7, 11.8, 11.9, and 11.10.
Hint: Use the Triangle Inequality for Proposition 11.6.

11.5 Let S' be the set of all points in Σ_2 that *don't* begin with 011. Use Definition 11.4a to prove that S' is open. Note that we already know that this is true since S' is the complement of the closed set S in Example 11.12b.

11.6 Let $r = 101010\ldots$ and $N_{\frac{1}{2}}(r) = \{s \mid d[s, r] < \frac{1}{2}\}$. Describe the set of points in $N_{\frac{1}{2}}(r)$.

11.7 The following problems complete the proof of Proposition 11.15.

a) Show that the shift map has 2^n periodic points with period n in Σ_2. Note that some of these points will not have prime period n.
Hint: Reviewing the proof of part (a) of Proposition 11.15 may help with this problem.

b) Show that the set of eventually periodic points that are not periodic is dense in Σ_2.
Hint: First, characterize the eventually fixed points. It may help to revisit the proof of part (a) of Proposition 11.15.

- c) In the proof of part (d) of Proposition 11.15, we claim that the orbit of the Morse sequence is dense in Σ_2. (The Morse sequence is defined in the proof.) Show that this is true. It may help to note that every finite sequence of 0's and 1's is eventually included in the Morse sequence.

d) In the proof of part (e) of Proposition 11.15, we assumed that the Morse sequence does not represent a periodic point in Σ_2. Why is this true?

11.8 Find all the points of period three in Σ_2. Which of these points have prime period three? Which are in the same orbit? Does Sarkovskii's Theorem apply here? Why or why not?

11.9 a) If $s = 101010\ldots$, then s is a period two point of σ. Determine the stable set of s.

b) Let t be any element of Σ_2. Describe the set of all points r in Σ_2 that satisfy the condition that $d[\sigma^n(r), \sigma^n(t)]$ approaches 0 as n goes to infinity.

11.10 Prove that ψ^{-1} as defined in Theorem 11.23 is continuous.

11.11 In the proof of Theorem 11.23, we defined the interval $I_{s_0 s_1 s_2}$ in Λ_2 for the point $s_0 s_1 s_2 \ldots$ in Σ_2. Let $s_0 s_1 s_2 \ldots$ and $t_0 t_1 t_2 \ldots$ be two points in Σ_2 that differ in at least one of the first three digits. Create an algorithm that will determine whether or not every point in $I_{s_0 s_1 s_2}$ is less than every point in $I_{t_0 t_1 t_2}$. (The intervals are disjoint, so one is "above" the other on the real line; the question is which one?)

11.12 Rewrite the statement of Theorem 9.3 so that it applies to metric spaces. Indicate which parts of the proof of Theorem 9.3 need to be changed so that it applies to the more general statement for metric spaces and explain what changes need to be made.

11.13 Let U and V be open subsets of the metric space X. Show that the set $U \cap V$ is also an open subset of X. (The fact that the intersection of open sets is open was used in the proof of Proposition 11.17.)

•11.14 THE DOUBLING MAP ON THE CIRCLE:
Let S^1 be the unit circle and identify each point on the circle by the radian measure of the angle between the positive x-axis and the ray beginning at the origin and passing through the point. We will always measure angles in a counterclockwise direction. The reader should recognize this as the usual representation of the unit

circle and angles on it. For example, the point $(1,0)$ is labeled 0, and the point $(0,1)$ is labeled $\frac{\pi}{2}$. We will also find it useful to identify the point α with the point $\alpha + 2n\pi$ where n is any integer. For example, the points 0, 2π, -2π, 4π, ... are all identical. Also, $\frac{\pi}{3}$, $\frac{7\pi}{3}$, and $-\frac{5\pi}{3}$ are all the same point on the circle. Again, this is the standard representation of angles on the unit circle.

We define a metric on S^1 by letting $d[\alpha, \beta]$ be the length of the shortest arc on the circle from α to β. More precisely, if α and β are in the interval $[0, 2\pi)$, then

$$d[\alpha, \beta] = \begin{cases} |\alpha - \beta| & \text{if } |\alpha - \beta| \leq \pi \\ |\alpha - \beta| - \pi & \text{if } |\alpha - \beta| > \pi. \end{cases}$$

Define the doubling function $\mathcal{D}: S^1 \to S^1$ by $\mathcal{D}(\theta) = 2\theta$.

a) Show that d is a metric for S^1.

b) Show that if $\alpha = \dfrac{\pi 2k}{2^n - 1}$ where k and n are natural numbers, then α is periodic with period n. Use this to show that periodic points of $\mathcal{D}$ are dense in S^1.

c) Let $(\alpha, \beta) = \{\theta \text{ in } S^1 \mid \alpha < \theta < \beta\}$ be any interval on S^1. Prove that there exists a natural number n such that $\mathcal{D}^n((\alpha, \beta)) = S^1$. Use this to show that $\mathcal{D}$ is topologically transitive on S^1.

d) Use Theorem 11.20 and the preceding parts of this problem to prove that $\mathcal{D}$ is chaotic on S^1.

e) Show that in every open interval on S^1 there exist points x_1 and x_2 and an integer n such that $d[\mathcal{D}^n(x_1), \mathcal{D}^n(x_2)] = \pi$.
    Hint: Consider points of the form $\dfrac{m\pi}{2^n}$.

f) Use part (e) and the definition of sensitive dependence to prove that $\mathcal{D}$ exhibits sensitive dependence on initial conditions. In particular, demonstrate that for every point x on S^1 and every $\epsilon > 0$ there is y on S^1 and a natural number n such that $d[x, y] < \epsilon$ and $d[\mathcal{D}^n(x), \mathcal{D}^n(y)] \geq \frac{\pi}{2}$.

11.15 **AN INVESTIGATION:** Let Υ be the set of all sequences in Σ_2 that contain no consecutive 1s.

a) Show that $\sigma(\Upsilon) = \Upsilon$.

b) How many periodic points does σ have in Υ?

c) Are the periodic points of σ dense in Υ?

d) Is $\sigma : \Upsilon \to \Upsilon$ chaotic?

11.16 Let D be a subset of a metric space and $f : D \to D$. Suppose that the periodic points of f are dense in D and there is a point whose orbit under iteration of f is dense in D.

a) Show that if D is finite, then D must consist of a single periodic orbit. Why does this imply that f is topologically transitive on D?
Hint: It may help to refer to the proof of Proposition 8.15.

b) Show that if D is infinite, then no iterate of any point with a dense orbit can be periodic. That is, if D is infinite and x_0 is either periodic or eventually periodic, then x_0 cannot have a dense orbit.

12
Newton's Method

The search for solutions of the equation $f(x) = 0$ is ancient, and methods that can solve equations of the form $ax^2 + bx + c = 0$ are several thousand years old. In the sixteenth century, Italian mathematicians discovered methods for solving third- and fourth-degree polynomials. However, it was shown in the early part of the nineteenth century that there is no general method for solving polynomials of degree five or higher. Consequently, methods for estimating solutions of equations as simple as polynomials are necessary. Isaac Newton developed such a method, which was later refined by Joseph Raphson, and which we now know as *Newton's method* or the *Newton–Raphson Method*. Newton's method is easy to use and is often taught in first semester calculus since it only requires knowledge of the derivative.

To illustrate Newton's method, suppose that $f(x)$ is differentiable and that x_0 is a reasonable approximation of a solution to the equation $f(x) = 0$. How can we find a better approximation? We begin by drawing the line tangent to the graph of $f(x)$ at the point $(x_0, f(x_0))$. The slope of this line is $f'(x_0)$. We label the point of intersection of the tangent line and the x-axis $(x_1, 0)$ and take x_1 as our next approximation. Since f is differentiable, its graph is reasonably straight on small intervals, and if x_0 is close to the solution, then x_1 is a better approximation than x_0. This procedure is illustrated in Figure 12.1, in which we we see that the new estimate of the solution x_1 is a significant improvement over the initial

128   12. Newton's Method

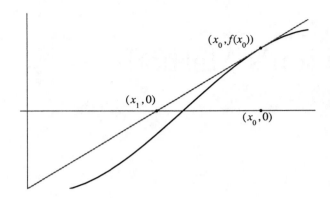

FIGURE 12.1. An illustration of Newton's method.

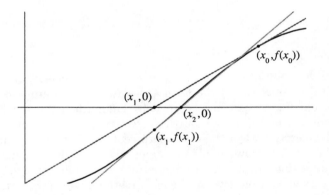

FIGURE 12.2. Two iterations of Newton's method.

estimate, x_0. In general, as long as the function doesn't curve too much between the estimate and the solution, the second estimate will be a better approximation than the first.

If we apply Newton's method again, we get an even better approximation, as shown in Figure 12.2. In fact, when Newton's method converges to a root, it does so at a rate that is proportional to the square of the error; the number of correct digits to the right of the decimal point approximately doubles at each iteration.

To develop an analytic understanding of Newton's method, suppose that we have a function $f(x)$, know its derivative $f'(x)$, and have x_0 as an initial estimate of a root. To find a second estimate x_1 in terms of x_0, we find an equation of the tangent line and solve for x when $y = 0$. As $(x_0, f(x_0))$ is a point on the line and the slope is $f'(x_0)$, the equation of the line is

$$(y - f(x_0)) = f'(x_0)(x - x_0).$$

Setting $y = 0$ and $x = x_1$ and solving for x_1, we find that

$$x_1 = x_0 - \frac{f(x_0)}{f'(x_0)}.$$

Now if we let x_1 be our estimate and apply the process again to find another estimate x_2, we see that

$$x_2 = x_1 - \frac{f(x_1)}{f'(x_1)}.$$

We can find another estimate and then another by repeatedly applying Newton's method. How can we represent this as an iterated function?

Define $N_f(x)$ by $N_f(x) = x - \frac{f(x)}{f'(x)}$. If x_0 is our initial estimate for Newton's method, then

$$x_1 = N_f(x_0)$$
$$x_2 = N_f(x_1) = N_f(N_f(x_0)) = N_f^2(x_0)$$

and, in general, if x_n is the nth estimate, then

$$x_n = N_f^n(x_0).$$

To simplify notation when the function in question is clear, we will suppress the subscript and write $N_f(x)$ as $N(x)$. We call $N_f(x)$ *Newton's function* for f.

EXAMPLE 12.1.
Let $p(x) = x^3 - x$. The roots of p are -1, 0, and 1. Let's see how fast we can find them using Newton's method. We begin by finding $N(x)$:

$$N(x) = x - \frac{p(x)}{p'(x)} = x - \frac{x^3 - x}{3x^2 - 1} = \frac{2x^3}{3x^2 - 1}.$$

If we estimate the root as .25, we get the following sequence of estimates:

$$x_0 = .25$$
$$x_1 = -.0385$$
$$x_2 = .000114$$
$$x_3 = -.00000000000299$$

The sequence converges quickly to the nearest root. If our initial value is .75, we get:

$$x_0 = .75$$
$$x_1 = 1.227$$
$$x_2 = 1.051$$
$$x_3 = 1.003$$

Again, the sequences converge quickly to the nearest root.

Now suppose that we try an initial value of .45. Iteration of Newton's method gives us the following sequence of values:

$$x_0 = .45$$
$$x_1 = -.464$$
$$x_2 = .567$$
$$x_3 = -10.156$$
$$x_4 = -6.793$$
$$x_5 = -4.562$$
$$x_6 = -3.091$$
$$x_7 = -2.135$$
$$x_8 = -1.536$$
$$x_9 = -1.192$$
$$x_{10} = -1.038$$
$$x_{11} = -1.002$$

It converges, but the convergence is not as quick and it is to the root that is furthest away from our initial estimate. This illustrates one of the problems with Newton's method. While it often finds a root, it may not find the closest one.

It is also worth noting that the values given for the previous calculations were rounded to the nearest thousandth after the calculation. The actual calculations were done with 16 significant digits. Given our previous work, we should not be too surprised to find out that one needs to be careful when rounding, as small errors can make a difference. For example, if we had started the last calculation at .46 rather than .45, then the sequence would have converged to 1. Similarly, had we rounded x_1 down to $-.47$, the sequence would again have converged to 1. To get some idea of the fragility of these calculations, the reader is encouraged to try other initial values for this function. Values between .4472 and .46 should prove to be especially interesting.

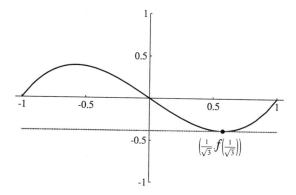

FIGURE 12.3. Newton's method applied to $p(x) = x^3 - x$ with the initial value $x_0 = \frac{1}{\sqrt{3}}$. Note that $p'(\frac{1}{\sqrt{3}})$ is zero, so the tangent line is horizontal, and we can't find x_1.

Finally, consider the initial value $\frac{1}{\sqrt{3}}$. Then $x_1 = N(\frac{1}{\sqrt{3}}) = \frac{2\sqrt{3}/9}{0}$, which, of course, is undefined. What happened? If we look at the graph of $p(x) = x^3 - x$ (Figure 12.3), we see that the tangent line at $(\frac{1}{\sqrt{3}}, p(\frac{1}{\sqrt{3}}))$ is horizontal. Consequently, the tangent line never intersects the x-axis and we can't find x_1. This manifests itself in Newton's method as a point for which $N(x) = x - \frac{p(x)}{p'(x)}$ is undefined since $p'(x) = 0$ there. We shall see later that this difficulty can be avoided for polynomials. □

The unpredictable behavior of Newton's method in the preceding example for initial values near .45 leads one to ask what else might go wrong. The answer is plenty. We will see that there are cubic polynomials and intervals of real numbers such that points in the intervals not only take a long time to converge but never converge to a root at all. However, we will also outline a procedure for determining intervals where the convergence is quick and to the nearest root in Theorem 12.15, and we will show in Theorem 12.2 that a number is an attracting fixed point of Newton's function for a polynomial if and only if it is a root of the polynomial.

THEOREM 12.2. *Let $p(x)$ be a polynomial. If we allow cancellation of common factors in the expression of $N_p(x)$, then $N_p(x)$ is always defined at roots of $p(x)$, a number is a fixed point of $N_p(x)$ if and only if it is a root of the polynomial, and all fixed points of $N_p(x)$ are attracting.*

One of the problems we wish to avoid with Theorem 12.2 is illustrated by the function $f(x) = x^2 - 2x + 1$. The only root of f is 1 and $f'(1) = 0$. As $N_f(x) = x - \frac{f(x)}{f'(x)}$, $N_f(1)$ is not defined since there is a zero in the denominator. More generally, if r is a root of $p(x)$ and $p'(r) = 0$, then $N_p(r) = r - \frac{0}{0}$, which is undefined. However, we see in the proof of the theorem that when the derivative at a root is 0, the numerator and denominator of the fraction $\frac{p(x)}{p'(x)}$ have common factors, which can be cancelled. The function $N_p(x)$ is defined at the root after cancellation of these factors.

PROOF. Let $p(x)$ be a polynomial and suppose that r is a root of p, that is, $p(r) = 0$. We recall from algebra that there is a natural number n and a polynomial q such that $p(x) = (x - r)^n q(x)$ and $q(r) \neq 0$. Hence,

$$\begin{aligned} N_p(x) &= x - \frac{p(x)}{p'(x)} \\ &= x - \frac{(x-r)^n q(x)}{(x-r)^n q'(x) + n(x-r)^{n-1} q(x)} \\ &= x - \frac{(x-r)^n q(x)}{(x-r)^{n-1}[(x-r)q'(x) + nq(x)]} \\ &= x - \frac{(x-r) q(x)}{(x-r)q'(x) + nq(x)}. \end{aligned}$$

Thus, if we cancel the term $(x-r)^{n-1}$, then $N_p(r)$ is defined and

$$N_p(r) = r - \frac{(r-r)q(r)}{(r-r)q'(r) + nq(r)} = r - \frac{0}{0 + nq(r)} = r$$

since $q(r) \neq 0$ by assumption. Therefore, $N_p(x)$ is defined at the root and has a fixed point there.

On the other hand, if $N_p(r) = r$, then

$$r - \frac{p(r)}{p'(r)} = r$$

where $\frac{p(x)}{p'(x)}$ is assumed to be in reduced form. This implies $\frac{p(r)}{p'(r)} = 0$ or $p(r) = 0$. Hence, fixed points of N_p are roots of p.

Further,

$$N'_p(x) = 1 - \frac{(p'(x))^2 - p(x)p''(x)}{(p'(x))^2} = \frac{p(x)p''(x)}{(p'(x))^2}. \qquad (12.1)$$

Thus, when r is a root of p and $p'(r) \neq 0$,

$$N_p(r) = r \quad \text{and} \quad |N'_p(r)| = \left|\frac{p(r)p''(r)}{((p'(r))^2}\right| = 0 < 1.$$

Therefore, r is an attracting fixed point of p. A similar argument holds when r is a root of p and $p'(r) = 0$. □

We should note that in the proof of Theorem 12.2 we only used the fact that the function was a polynomial to guarantee that $N_p(x)$ was defined at the root. The same proof demonstrates that when r is a root of the function f and $N_f(r)$ is defined, then $N_f(r) = r$ and $N_f'(r) = 0$. Fixed points whose derivatives are zero are often called *super attracting* fixed points. We will investigate the significance of this designation in the exercises. As a corollary to the proof of Theorem 12.2 we get the following.

COROLLARY 12.3. *If f is a differentiable function, then a fixed point of $N_f(x)$ must be an attracting fixed point of N_f and a root of f. On the other hand, if a point is a root of f and N_f is defined at that point, then the point is an attracting fixed point of N_f.*

12.1. Newton's Method for Quadratic Functions

Before we look at some of the strange behaviors associated with Newton's method, let's look at a case where things go right. Let $f(x) = ax^2 + bx + c$ where $a \neq 0$. Then

$$N_f(x) = x - \frac{ax^2 + bx + c}{2ax + b} = \frac{ax^2 - c}{2ax + b}. \quad (12.2)$$

In the next proposition, we will show that if we are interested in analyzing the dynamics of Newton's method, then we may assume $a = 1$ and $b = 0$.

PROPOSITION 12.4. *Let $f(x) = ax^2 + bx + c$ and $q(x) = x^2 - A$ where $A = (b^2 - 4ac)$. Then $\tau(x) = 2ax + b$ is a topological conjugacy from $N_f(x)$ to $N_q(x)$.*

PROOF. In Exercise 2.15, we proved that all nonconstant linear functions are homeomorphisms. Thus, we may assume that $\tau(x) = 2ax + b$ is a homeomorphism. It remains to show that $\tau \circ N_f = N_q \circ \tau$. From equation (12.2), we have $N_f(x) = \dfrac{ax^2 - c}{2ax + b}$ and $N_q(x) = \dfrac{x^2 + A}{2x}$. Then

$$(\tau \circ N_f)(x) = 2a \left(\frac{ax^2 - c}{2ax + b} \right) + b = \frac{2a^2 x^2 + 2abx + b^2 - 2ac}{2ax + b}$$

and

$$(N_q \circ \tau)(x) = \frac{(2ax + b)^2 + A}{2(2ax + b)} = \frac{4a^2 x^2 + 4abx + b^2 + A}{2(2ax + b)}$$
$$= \frac{4a^2 x^2 + 4abx + b^2 + b^2 - 4ac}{2(2ax + b)} = \frac{2a^2 x^2 + 2abx + b^2 - 2ac}{2ax + b}.$$

Thus, $\tau \circ N_f = N_g \circ \tau$, and the proof is complete. □

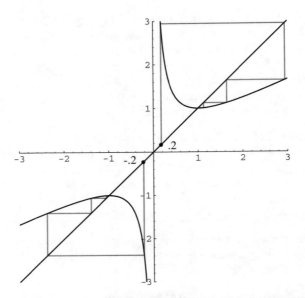

FIGURE 12.4. Graphical analysis for Newton's method applied to $p(x) = x^2 - 1$ with .2 and $-.2$ used as starting points.

As a consequence of Proposition 12.4, we may restrict our analysis of Newton's method for quadratic polynomials to polynomials of the form $q(x) = x^2 - c$. Obviously, when $c > 0$ the solutions of $q(x) = 0$ are $\pm\sqrt{c}$, when $c = 0$ the only solution is 0, and when $c < 0$ there are no real solutions. We begin our analysis of the case $c > 0$ with an example.

EXAMPLE 12.5.
Let $p(x) = x^2 - 1$. From Theorem 12.2, we know that -1 and 1 are the only fixed points of $N(x)$. Since $p'(0) = 0$, $N(0)$ is undefined. We will show that $W^s(1) = (0, \infty)$ and $W^s(-1) = (-\infty, 0)$. Graphical analysis of $N(x)$ justifies this claim, though of course it doesn't prove it. The graph of $N(x)$ with graphical analysis done for two initial points is shown in Figure 12.4.

We begin by proving the claim for $W^s(1)$. We note that if x_0 is in $(0, \infty)$, then $N(x_0) = \frac{x_0^2 + 1}{2x_0}$ is in $[1, \infty)$ since this is equivalent to saying that $x_0^2 - 2x_0 + 1 \geq 0$. It remains to show that $\lim_{n \to \infty} N^n(x_1) = 1$ whenever $x_1 > 1$.

Let $x_1 > 1$. Equation (12.1), implies that $N'(x) = \frac{q(x)q''(x)}{(q'(x))^2} = \frac{x^2 - 1}{2x^2}$. Differentiating again, we find that $N''(x) = \frac{1}{x^3}$. Since $N''(x) > 0$ when x is in the interval $(1, \infty)$, we know that $N'(x)$ is increasing on this interval.

As $N'(1) = 0$ and
$$\lim_{x \to \infty} N'(x) = \lim_{x \to \infty} \frac{x^2 - 1}{2x^2} = \frac{1}{2},$$
this implies that $0 < N'(x) < \frac{1}{2}$ for all x in $(1, \infty)$. Now, the Mean Value Theorem implies that there is c_1 between 1 and x_1 such that
$$N(x_1) - 1 = N(x_1) - N(1) = (N'(c_1))(x_1 - 1).$$
Since $0 < N'(x) < \frac{1}{2}$ for all x in $(1, \infty)$, we see that
$$0 < N(x_1) - 1 < \tfrac{1}{2}(x_1 - 1).$$
Thus, $N(x_1) > 1$ and $|N(x_1) - 1| < \frac{1}{2}|x_1 - 1|$. Since $N(x_1) > 1$, we can continue by induction to find a sequence of numbers c_n between $N^{n-1}(x_1)$ and 1 so that
$$\begin{aligned} |N^n(x_1) - 1| &= |N(N^{n-1}(x_1)) - N(N^{n-1}(1))| \\ &= |f'(c_n)||N^{n-1}(x_1) - N^{n-1}(1)| \\ &< \left(\tfrac{1}{2}\right)\left(\tfrac{1}{2}\right)^{n-1}|x_1 - 1| = \left(\tfrac{1}{2}\right)^n |x_1 - 1|. \end{aligned}$$
The last term in the previous inequality approaches 0 as n tends to infinity, so it follows that if $x_1 > 1$, then $N^n(x_1)$ converges to 1 as n tends to infinity. Therefore, if $x_0 > 0$, then $N^n(x_0)$ also converges to 1 as n tends to infinity. Thus, we have shown that $(0, \infty) \subset W^s(1)$.

The proof that $(-\infty, 0) \subset W^s(-1)$ is similar. As these two inclusions account for all the points in the domain of N, the conclusion $W^s(1) = (0, \infty)$ and $W^s(-1) = (-\infty, 0)$ follows. □

The argument in the preceding example suggests the following proposition. The proof of the proposition is left as an exercise. Two possible approaches to the proof are suggested in Exercise 12.18.

PROPOSITION 12.6. *If $q(x) = x^2 - c^2$ and $c > 0$, then $N_q(x)$ has fixed points at $\pm c$, both of which are attracting. Further, the stable set of c is the interval $(0, \infty)$, and the stable set of $-c$ is the interval $(-\infty, 0)$.*

The case $c = 0$ is easier. In this case, our function is $q(x) = x^2$ and $N(x) = x - \frac{x^2}{2x} = \frac{1}{2}x$. Demonstrating that 0 is the only fixed point of $N(x)$ and that the stable set of 0 is the set of all real numbers is a straightforward exercise.

Now we turn to the more interesting case, Newton's method for the function $q(x) = x^2 + c$ when $c > 0$. Of course $q(x)$ doesn't have any real roots and, as the only attracting fixed points of N are roots of q, we expect something different to happen. It is easiest to first consider the special case $c = 1$.

EXAMPLE 12.7.

Let $p(x) = x^2 + 1$. Then $N(x) = \frac{x^2-1}{2x}$. To get some idea of what might happen for this function we look at graphical analysis for $N(x)$ with initial point .1 in Figure 12.5.

Obviously this isn't as simple as the cases we have examined already. Looking at Figure 12.5, we are led to wonder if the action of this function might, in fact, be chaotic. We prove that this is the case by showing that N is related to the doubling map of the circle.

Recall from Exercise 11.14 that the circle is denoted S^1 and is usually depicted as the unit circle. Angles or points on the circle are measured in radians in the counterclockwise direction with the initial point on the positive x-axis. The doubling map, $\mathcal{D}: S^1 \to S^1$, is defined by $\mathcal{D}(\theta) = 2\theta$. We demonstrated in Exercise 11.14 that $\mathcal{D}$ was chaotic on S^1.

We define a map from the circle with the point 0 deleted to the real numbers by $\phi(x) = \cot(\frac{x}{2})$. This map satisfies $N \circ \phi = \phi \circ \mathcal{D}$, since by the half angle identities,

$$(N \circ \phi)(x) = \frac{\cot^2\left(\frac{x}{2}\right) - 1}{2\cot\left(\frac{x}{2}\right)} = \frac{\cos^2\left(\frac{x}{2}\right) - \sin^2\left(\frac{x}{2}\right)}{2\cos\left(\frac{x}{2}\right)\sin\left(\frac{x}{2}\right)}$$
$$= \frac{\cos(x)}{\sin(x)} = \cot(x) = (\phi \circ \mathcal{D})(x).$$

Note that ϕ is not defined at 0 and so is not a homeomorphism of S^1 and $\mathbb{R}$. Hence, it is not a topological conjugacy. We leave as an exercise the verification of the fact that $\cot(\frac{x}{2})$ is continuous, one-to-one, and onto when considered as a map from the circle with the point 0 deleted to the real numbers. We also note that $(\phi \circ \mathcal{D})(\pi) = \phi(0)$, which is not defined. However, $(N \circ \phi)(\pi) = N(0)$, which is not defined either since $f'(0) = 0$. Working backwards, we see that $\mathcal{D}^2(\pi/2) = \mathcal{D}^2(3\pi/2) = 0$ and $\phi \circ \mathcal{D}^2$ is not defined at $\pi/2$ or $3\pi/2$. On the other hand, both $(N^2 \circ \phi)(\pi/2)$ and $(N^2 \circ \phi)(3\pi/2)$ reduce to $N(0)$, which is not defined either. Iterating this procedure, we can show that r_n is a point on S^1 for which $\mathcal{D}^n(r_n) = 0$ and $(\phi \circ \mathcal{D}^n)(r_n)$ is not defined if and only if $N^n(\phi(r_n))$ is not defined. So, ϕ always maps points in S^1 that are not mapped to 0 by some iterate of $\mathcal{D}$ to points in $\mathbb{R}$ for which N^n is defined for all natural numbers n. Also, if γ is a point in S^1 that is not mapped to 0 by an iterate of $\mathcal{D}$, then $(\phi \circ \mathcal{D}^n)(\gamma) = (N^n \circ \phi)(\gamma)$ for all natural numbers n.

By Theorem 8.16, it suffices to show that periodic points of N are dense in $\mathbb{R}$ and that N is topologically transitive on R to demonstrate that N is chaotic on $\mathbb{R}$. Let (a, b) be an interval in R. Then $\phi^{-1}((a,b)) = (\alpha, \beta)$ is an interval in S^1. By Exercise 11.14b, there is γ in (α, β) and k such that $\mathcal{D}^k(\gamma) = \gamma$. Without loss of generality, we can assume that $\mathcal{D}^n(\gamma) \neq 0$

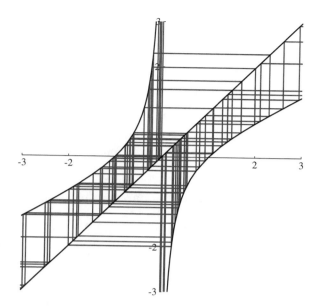

FIGURE 12.5. Graphical analysis for 100 iterations of Newton's function for $p(x) = x^2 + 1$ with .1 used as a starting point.

for any n. (Why?) Hence, if $c = \phi(\gamma)$, then c is in (a, b) and

$$N^k(c) = N^k(\phi(\gamma)) = \phi(\mathcal{D}^k(\gamma)) = \phi(\gamma) = c.$$

Thus, c is a periodic point of N in (a, b), and the periodic points of N are dense in $\mathbb{R}$. Also, we demonstrated in Exercise 11.14c that there exists m such that $\mathcal{D}^m((\alpha, \beta)) = S^1$. It follows that $N^m((a, b)) = \mathbb{R}$. Therefore, if (d, e) is any other interval in $\mathbb{R}$, then there is x_0 in (a, b) such that $N^m(x_0)$ is in (d, e) and N is topologically transitive.

A slightly different proof that N is chaotic on the set of real numbers is outlined in Exercise 12.16, and a proof using complex analysis is outlined in Example 14.21. □

Now suppose that $q(x) = x^2 + c^2$, $c \neq 0$, and $p(x) = x^2 + 1$. By showing that N_q and N_p are topologically conjugate we are able to prove the following proposition. The details of the proof are left as an exercise.

PROPOSITION 12.8. *If $q(x) = x^2 + c^2$, then N_q is chaotic on the set $(-\infty, 0) \cup (0, \infty)$.*

In Chapter 14, we will extend our discussion by investigating the dynamics of Newton's method when applied to complex quadratic functions.

12.2. Newton's Method for Cubic Functions

In this section, we will find that the behavior of Newton's method for a cubic polynomial is much richer than its quadratic cousin. As with the quadratic, we can simplify the analysis by reducing to cases where the behavior of N varies with a single parameter.

PROPOSITION 12.9. *If $f(x) = ax^3 + bx^2 + cx + d$ and $g(x) = x^3 + Ax + B$ where $A = 9ac - 3b^2$ and $B = 27a^2d + 2b^3 - 9abc$, then the function $\tau(x) = 3ax + b$ is a topological conjugacy from $N_f(x)$ to $N_g(x)$. That is, $\tau \circ N_f = N_g \circ \tau$.*

The proof of Proposition 12.9 is an exercise in algebra and is left to the reader. More interesting questions are: Why should we expect there to be a linear function that conjugates N_f and N_g? What motivates the choice of $\tau(x) = 3ax + b$ as the conjugacy? The answer to the second question is that τ maps the inflection point of f to the inflection point of g. If we can understand why that is the proper point to consider, we will understand the answer to the first question. It is really a matter of translation and scaling. The placement of the y-axis relative to the graph of the function is irrelevant to the behavior of Newton's method for the function. Hence, we can translate horizontally. Furthermore, if we divide $f(x) = ax^3 + bx^2 + cx + d$ by a to get $p(x) = x^3 + \frac{b}{a}x^2 + \frac{c}{a}x + \frac{d}{a}$, we find that $N_f = N_p$. In other words, we can rescale f by a. We leave the proof of this fact as an exercise.

PROPOSITION 12.10. *Let $f(x) = x^3 + ax + b^3$ where $b \neq 0$ and define $g(x) = x^3 + cx + 1$ where $c = \frac{a}{b^2}$. Then N_f and N_g are topologically conjugate via the homeomorphism $\tau(x) = \frac{1}{b}x$.*

Proposition 12.10 coupled with Proposition 12.9 implies that either the dynamics of Newton's method for a cubic polynomial are the same as the dynamics of Newton's method for a polynomial in the parametrized family $f_c(x) = x^3 + cx + 1$ or the dynamics are the same as those of Newton's method for a polynomial of the form $g_a(x) = x^3 + ax$. The following proposition implies that Newton's method for functions of the latter form must be topologically conjugate to Newton's method for one of three polynomials:

$$p_+(x) = x^3 + x, \qquad p_-(x) = x^3 - x, \qquad \text{or} \qquad p_0(x) = x^3.$$

PROPOSITION 12.11. *If $\tau(x) = \frac{1}{a}x$ where $a \neq 0$, $g(x) = x^3 + a^2x$, and $p_+(x) = x^3 + x$, then N_g and N_{p_+} are topologically conjugate via τ. On the other hand, if $f(x) = x^3 - a^2x$ and $p_-(x) = x^3 - x$, then τ is a topological conjugacy between N_g and N_{p_-}.*

Using this proposition and the preceding discussion, we conclude that for all cubic polynomials $p(x)$, N_p is topologically conjugate to N_f where $f(x) = x^3 + cx + 1$ or N_p is topologically conjugate to Newton's method for one of the three polynomials, $p_+(x) = x^3 + x$, $p_-(x) = x^3 - x$, or $p_0(x) = x^3$. The analysis of N_{p_+} and N_{p_0} is easy and is left as an exercise. The results are summarized in the following proposition.

PROPOSITION 12.12. *Let $p_+(x) = x^3 + x$ and $p_0(x) = x^3$. Then 0 is the only fixed point of both N_{p_+} and N_{p_0}, and the stable set of 0 is $\mathbb{R}$ for both functions.*

The behavior of Newton's method for $p_-(x) = x^3 - x$ is more complicated. We considered some of the issues in Example 12.1. Additional interesting behavior will be examined in Exercise 12.5 at the end of this chapter.

We turn now to Newton's method for $f(x) = x^3 + cx + 1$. In this case

$$N(x) = \frac{2x^3 - 1}{3x^2 + c}.$$

When $c > 0$, it is easy to verify that N has exactly one root and that the stable set of the root contains all of $\mathbb{R}$. When $c = 0$, there is still one real root and the stable set contains all real numbers except 0.

Now consider the case $c < 0$. We note that the graph of f is symmetric about the point $(0,1)$. Also, f has a relative minimum at $x = \sqrt{-c/3}$ and a relative maximum at $x = -\sqrt{-c/3}$. We label the x value at the minimum point x_m. The graph of f is shown in Figure 12.6 along with the graph of N for two values of c.

We note that as c decreases, the relative minimum of f decreases and the relative maximum increases. When $c = -\frac{3}{2}(2^{1/3})$, or about -1.8898, the relative minimum is 0 and f has two real roots, one at 0 and the other at $x = x_m$. We denote the point $-\frac{3}{2}(2^{1/3})$ by c_0. Note that when c is less than c_0, f has three real roots. When c is greater than c_0, f has one real root. Looking at Figure 12.6, we ask what happens to Newton's method when c is between 0 and c_0. For example, suppose that $c = -1$. If we start Newton's method at 0, then the next value is larger than x_m. The value after that seems to be smaller than x_m and then it becomes hard to discern. Iterating on a computer, we find that when $c = -1$ we get the sequence

$$x_0 = 0, \quad x_1 = 1, \quad x_2 = .5, \quad x_3 = 3, \quad x_4 = 2.038, \quad x_5 = 1.39,$$
$$x_6 = .912, \quad x_7 = .345, \quad x_8 = 1.428, \quad x_9 = .942, \quad x_{10} = .405$$

It doesn't look like this sequence will converge to the root anytime soon.

140   12. Newton's Method

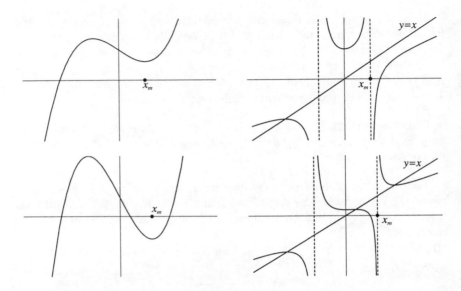

FIGURE 12.6. The graphs of $f(x) = x^3 + cx + 1$ and N_f are shown in the first row for the parameter value $c = -1$. In the second row, we again see f and N_f, this time for the parameter value $c = -3.2$. The vertical lines in the graph of N_f are the asymptotes created at the points where $f'(x) = 0$.

To investigate this behavior further, we graph the first 100 iterates of 0 for a variety of parameter values between -2 and 0 in Figure 12.7. Note that 0 seems to converge to a positive root when such a root exists, that is, when $c < c_0 \approx -1.89$. The dark line running slightly upward across the bottom of the diagram when c is larger than c_0 represents the negative root, so we see that on a fairly regular basis if $c > c_0$, then 0 is attracted to the negative root within 100 iterations of N_c. If 0 is not attracted quickly to one of the roots, then there does not seem to be any apparent pattern to its behavior under iteration of N_c. If we look at the 500th through 600th iterates of 0, we get a clearer picture, since by that time the iterates should be near an attracting orbit. The results of such a computation are shown in Figure 12.8. Again, there is no obvious pattern in the behavior of N_c at those parameter values for which 0 does not converge to a root, though a large percentage of the orbits do seem to be converging to a root.

Looking at Figure 12.8, we wonder what is happening for those parameter values where Newton's method isn't converging to the root quickly. Note that when $c \approx -1.265$ the orbit of 0 tends to clump in two places,

12.2. Newton's Method for Cubic Functions

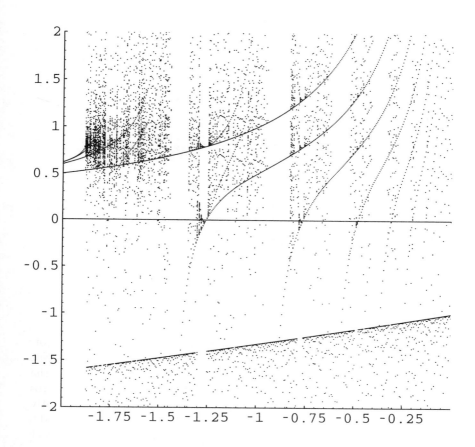

FIGURE 12.7. The points $(c, N_c^n(0))$ are plotted where N_c is Newton's function for $f_c(x) = x^3 + cx + 1$. We use 500 parameter values of c between -2 and 0, and n varies from 0 to 100 for each c.

142   12. Newton's Method

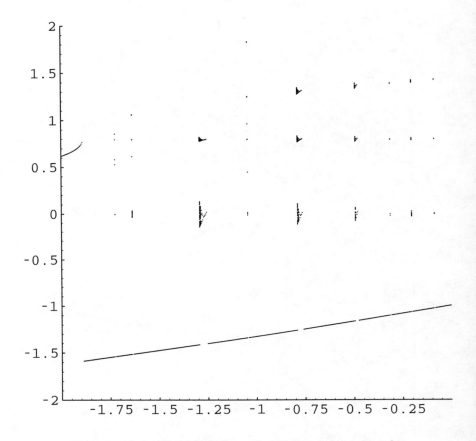

FIGURE 12.8. The 500th through 600th points in the orbit of 0 under iteration of Newton's function for the cubic polynomial $f(x) = x^3 + c + 1$ are shown. We use 500 parameter values between -2 and 0.

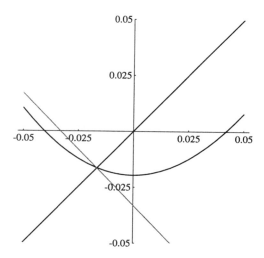

FIGURE 12.9. The graphs of $N_f^2(x)$ and the line $y = x$ where $f(x) = x^3 - 1.265x + 1$. Note that the graph of N_f^2 crosses the line $y = x$ near the point $(-.016, -.016)$. Note also that the slope of N_f at the intersection is negative and greater than -1. A light line with slope -1 is drawn through the point of intersection.

one near 0 and the other near .8. Consequently, in the next example, we continue our study of Newton's method by investigating N for $c = -1.265$.

EXAMPLE 12.13.

Let $f(x) = x^3 - 1.265x + 1$. Then $N(x) = \frac{2x^3 - 1}{3x^2 - 1.265}$. When iterating N starting at 0, we get the following sequence:

$$x_0 = 0, \quad x_1 = .79051, \quad x_2 = -.019675, \quad x_3 = .79125,$$
$$x_4 = -.015043, \quad x_5 = .79094, x_6 = -.016973, \quad x_7 = .79106,$$
$$x_8 = -.016232, \quad x_9 = .79101, \quad x_{10} = -.016527$$

We seem to be converging to an attracting periodic point with prime period two!

To investigate further, we graph $N^2(x)$ in a neighborhood around 0 in Figure 12.9 If there is an attracting periodic point with period two in this neighborhood, then $N^2(x)$ has an attracting fixed point there. In this case, the graph of $N^2(x)$ will intersect the line $y = x$ and the slope of the graph of $N^2(x)$ will be less than one in absolute value at the point of intersection.

144   12. Newton's Method

While Figure 12.9 presents powerful evidence that there is an attracting period two point of N near $-.016$, it is a fact that should still be proven. To do this, we look for an interval I such that the intersection of $N(I)$ and I is empty, $N^2 : I \to I$, and $|(N^2)'(x)| < 1$ for all x in I. The first condition ensures that I contains no fixed point of N, and Theorem 6.3, along with the last two conditions, guarantees that N has an attracting periodic point with prime period two in I. Figure 12.9 suggests that the interval we are looking for might be $[-.03, .03]$. Our goal then is to show that $I = [-.03, .03]$ satisfies the following three properties:

(1) $N(I) \cap I = \emptyset$,
(2) $N^2 : I \to I$, and
(3) $|(N^2)'(x)| < 1$ for all x in I.

In equation (12.1), we calculated that $N_f'(x) = \frac{f(x)f''(x)}{(f'(x))^2}$. Hence, for $f(x) = x^3 - 1.265x + 1$, we have $N'(x) = \frac{(x^3 - 1.265x + 1)(6x)}{(3x^2 - 1.265)^2}$. Differentiating and simplifying, we find that

$$N''(x) = \frac{6(2(1.265)x^3 - 9x^2 + 2(1.265)^2 x - 1.265)}{(3x^2 - 1.265)^3}$$

$$= \frac{6x^2[2.53x - 9] + 6(1.265)[2.53x - 1]}{(3x^2 - 1.265)^3}. \tag{12.3}$$

For x in the interval $I = [-.03, .03]$, we note that both the numerator and the denominator in $N''(x)$ are negative. Consequently, $N''(x)$ is positive and N' is increasing on the interval. That is,

$$-.12 < N'(-.03) \leq N'(x) \leq N'(0) = 0 \tag{12.4}$$

for x in $[-.03, 0]$ and

$$0 = N'(0) \leq N'(x) \leq N'(.03) < .11 \tag{12.5}$$

for x in $[0, .03]$. Hence, N is decreasing on $[-.03, 0]$ and increasing on $[0, .03]$. Since $N(-.03) \approx .7922$, $N(0) \approx .7905$, and $N(.03) \approx .7922$, this implies that $N(I)$ is contained in the interval $[.79, .793]$. Clearly, $N(I) \cap I = \emptyset$ and we have demonstrated condition 1.

Using equation (12.3) again, we find that $N''(x) < 0$ on $[.79, .793]$. Thus, N' is decreasing on the interval $[.79, .793]$, and if x is in $[.79, .793]$, then

$$6.4 > N'(.79) \geq N'(x) \geq N'(.793) > 6.1. \tag{12.6}$$

This implies that N is increasing on $[.79, .793]$. Since $N(.79) \approx -.0229$ and $N(.793) \approx -.00426$, we conclude that $N([.79, .793]) \subset [-.03, .03]$ and $N^2 : I \to I$, as required by condition 2.

Finally, we wish to estimate $(N^2)'(x)$ for x in I. From Exercise 6.2, we know that $(N^2)'(x) = N'(x)N'(N(x))$. Since inequalities (12.4) and (12.5) imply $|N'(x)| < .12$ when x in I and inequality (12.6) implies $|N'(x)| < 6.4$ when x is in $[.79, .793]$, we have

$$|(N^2)'(x)| = |N'(x)||N'(N(x))| < (.12)(6.4) = .768 < 1,$$

as required by condition 3.

We have shown that N has an attracting periodic point with prime period two in $[-.03, .03]$. We leave as an exercise the proof that $[-.03, .03]$ is contained in the stable set of this periodic point. □

In the preceding example, we found that when $p(x) = x^3 - 1.265x + 1$, there are intervals on which N not only doesn't converge to a root but converges to an attracting periodic point instead. Given the size of the clumps in Figure 12.8 for parameter values near -1.265, we are led to wonder if we won't find something more interesting than a period two point. In Figure 12.10, we investigate by plotting an enlargement of the lower of the two clumps.

Amazingly, we find the familiar bifurcation diagram of Figure 10.1. In the exercises, we will search for an explanation of this phenomenon. The diagram indicates that N^2 goes through a cascade of period-doubling bifurcations as the parameter decreases from -1.25 to -1.3. It also indicates something more is going on. Recall that in the case of the logistic equation, the cascade of period-doubling bifurcations was followed by chaos. So, for the cubic case, we are led to the following conjecture.

CONJECTURE 12.14. *If $f(x) = x^3 - cx + 1$, then there exists $c > 0$ and a set of real numbers on which N_f is chaotic. Further, this set is attracting.*

We will explore this conjecture in the exercises. For another discussion of the dynamics of Newton's method for cubic polynomials, we refer the reader to the article of that name by J. Walsh, which can be found in the January 1995 issue of the *College Math Journal*.

12.3. Intervals and Rates of Convergence

The results of the previous section on Newton's method for cubic polynomials should have us a bit concerned by now. After all we have shown that if we do not choose our initial guess wisely, then not only might the method not converge, but it might wander periodically, or perhaps even chaotically, forever. Fortunately, there are tests to determine whether or not our initial guess converges to the closest root. Theorem 12.15 outlines one such criterion. One might also wonder why we would choose to use a method that is subject to problems like those just demonstrated. The reason is

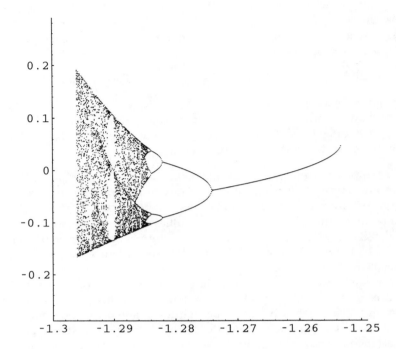

FIGURE 12.10. The graph of the 500th through 600th points in the orbit of zero under iteration of Newton's function for $f(x) = x^3 + cx + 1$ for parameter values between -1.3 and -1.25. The parameter is graphed on the horizontal axis and 400 values are used.

speed. Newton's method exhibits quadratic convergence. Essentially, this means that the number of correct digits to the right of the decimal place in our estimate doubles with each iteration of Newton's method. Consequently, if we can find an approximation that is reasonably close to the root, then we can get a very good approximation of the root with only a few steps. By contrast, the bisection method, which is also commonly taught in calculus or precalculus classes, converges geometrically. That is, it takes approximately three iterations to add one more correct digit to the approximation. The rate of convergence of Newton's method is made more precise in Theorem 12.15.

THEOREM 12.15. *Suppose that $f(x)$ is a twice differentiable function and $f'(x_0)$ is not equal to 0. Let*

$$\epsilon = \left|\frac{f(x_0)}{f'(x_0)}\right|, \qquad x_1 = N_f(x_0), \qquad and \qquad I = [x_1 - \epsilon, x_1 + \epsilon].$$

If $|f''(x)| < \frac{|f'(x_0)|^2}{2|f(x_0)|}$ for all x in I, then $N_f^n(x_0)$ converges to a root of f in I and

$$\left|N_f^{n+1}(x_0) - N_f^n(x_0)\right| \leq \frac{1}{\epsilon} \cdot \left|N_f^n(x_0) - N_f^{n-1}(x_0)\right|^2.$$

As one might guess, Theorem 12.15 is proven by induction on the number of iterations. While the proof requires no mathematical concepts other than induction and first-year calculus, a proper treatment of it requires several pages. Inclusion of the proof here will not serve to increase our understanding of dynamical systems, so we refer the interested reader to the text by Ostrowskii, which is listed in the references.

Exercise Set 12

12.1 For each of the following functions, use Newton's method to find as many roots of the equation as you can. Describe the stable set of each root. Indicate any points you find that are not in the stable set of any root and explain why they aren't in the stable set of a root.

a) $f(x) = x^2 - 3x + 2$

b) $g(x) = \log x$

c) $s(x) = \sec x$

d) $E(x) = e^x$

e) $k(x) = \frac{1}{x} - 1$

f) $p(x) = x^4 + x^2$

g) $r(x) = \sqrt{x}$

h) $F(x) = \frac{x}{x^2 + 1}$

12.2 Draw the graph of a function for which Newton's method converges to infinity whenever the initial value is greater than 2 and for which it converges to 0 whenever the initial value is less than 2. Find a formula for such a function. Prove that your formula works; that is, show that the stable sets of zero and infinity are as desired.

12.3 Investigate Newton's method for $f(x) = x^2$. Find the unique fixed point and prove that its stable set is the set of all real numbers.

12.4 If p is a fixed point of the function f, then p is said to be super attracting if $f'(p) = 0$. Prove that if p is a super attracting fixed point of f, then for each $\epsilon > 0$, there is $\delta > 0$ such that if $|p-y| < \delta$, then $|f(y) - p| < \epsilon|p - y|$. Explain why this property suggests the name super attracting.

12.5 NEWTON'S METHOD FOR $p(x) = x^3 - x$:
Suppose that $p(x) = x^3 - x$. In Example 12.1, we demonstrated that $N(x) = \frac{2x^3}{3x^2-1}$. As we noted in that example, $N(x)$ is undefined for $x = \pm\frac{1}{\sqrt{3}}$.

a) Show that $W^s(0) = (-\frac{1}{\sqrt{5}}, \frac{1}{\sqrt{5}})$. What happens to $\frac{1}{\sqrt{5}}$?

b) Find a repelling periodic point of N with period two.

c) Show that there exists a sequence of points $a_0, a_1, a_2, \ldots$ such that $a_0 = \frac{1}{\sqrt{3}}$, $a_i > a_{i+1}$ for all natural numbers i, $N^i(a_i) = \frac{1}{\sqrt{3}}$ if i is even and $N^i(a_i) = -\frac{1}{\sqrt{3}}$ if i is odd.

Show that $\lim_{i \to \infty} a_i = \frac{1}{\sqrt{5}}$.

What happens to Newton's method if our initial point is one of the numbers a_i?

If we numerically solve for a_2 and use a computer to apply Newton's method to this numerical solution, then it converges to one of the roots -1 or 1. Why don't we see the behavior that our analysis predicts?

Demonstrate the existence of a similar sequence $b_0, b_1, b_2, \ldots$, which begins at $-\frac{1}{\sqrt{3}}$ and converges to $-\frac{1}{\sqrt{5}}$.

d) Show that (a_{i+1}, a_i) is in $W^s(1)$ whenever i is odd and in $W^s(-1)$ whenever i is even. State and prove a similar fact for the intervals (b_i, b_{i+1}).

e) Summarize by classifying the real numbers by their behavior under iteration of N.

12.6 AN INVESTIGATION: Consider the sequence $d_i = \dfrac{a_{i-1} - a_i}{a_i - a_{i+1}}$ where the a_i are as defined in Exercise 12.5. Use a computer to determine numerically that d_i converges. To what number does it converge? Why is this so? Can you prove that your answer is correct?

12.7 Prove Proposition 12.12. Let $p_+(x) = x^3 + x$ and $p_0(x) = x^3$. Show that the only fixed point of Newton's method for both functions is 0 and that $W^s(0)$ is $\mathbb{R}$.

12.8 Let $f(x) = x^3 + cx + 1$ where $c > 0$. Show that N_f has exactly one fixed point and that the stable set of the fixed point contains all real numbers. Show that when $c = 0$ there is still only one fixed point, and its stable set contains all real numbers except 0.

12.9 Let $f(x) = x^3 - 1.265x + 1$. Show that the interval $[-.03, .03]$ is in the stable set of an attracting periodic point of N_f.
Hint: Use the results of Example 12.13.

12.10 In Example 12.13, we stated that if $N(x)$ has an attracting periodic point of period two in an interval, then the graph of $N(x)$ crosses the line $y = x$ in the interval and that $|N'(x)| < 1$ at the point of intersection. Explain why this is so.

12.11 Let $f(x) = x^3 + cx + 1$. Find a value of the parameter c for which N_f has an attracting periodic point with prime period four.
Hint: Use the bifurcation diagram in Figure 12.10.

12.12 a) Figure 12.10 indicates that Newton's method for the function $f(x) = x^3 + cx + 1$ goes through a cascade of period-doubling bifurcations as c varies from -1.25 to -1.3. Use a sequence of graphs of N^2 for parameter values in this range to explain why this is to be expected. Why do we use the graph of N^2 instead of the graph of N?

b) Using the information from part (a) and our experience with period-doubling cascades in Chapter 11, predict a parameter value and an interval on which we would expect N^2 to be chaotic. Explain why this is a good choice. Can you find numerical or graphical evidence to support your guess? Can you prove that your guess is correct?

c) If N^2 is chaotic on I, does it follow that N is chaotic on $I \cup N(I)$? Explain.

12.13 Let $f(x) = x^3 + cx + 1$. Find a value of c for which N_f has an attracting periodic point with prime period three.

Hint: Examination of Figure 12.7 should yield some likely candidates. Note that in Example 12.13 we found a period two point for $c = -1.265$ at which value we see two "blobs" in the bifurcation diagram.

*12.14 Let $f(x) = x^3 + cx + 1$. Find a value of c for which there is an interval I in which N_f has periodic points of all prime periods.

12.15 Let $\phi : S^1 \setminus \{0\} \to \mathbb{R}$ be as defined in Example 12.1.

a) Show that ϕ is one-to-one, onto, and continuous.

b) Show that if γ is a point in S^1 that is not mapped to 0 by an iterate of D, then $(\phi \circ D^n)(\gamma) = (N^n \circ \phi)(\gamma)$ for all natural numbers n.

c) Let (a, b) be an interval in $\mathbb{R}$ and (α, β) be $\phi^{-1}((a, b))$. Suppose that $D^m((\alpha, \beta)) = S^1$. Prove that $N^m((a, b)) = \mathbb{R}$.

12.16 Let S^* be the set of points of S^1 that are not eventually fixed at 0 by the doubling map $D(\theta) = 2\theta$. That is,

$$S^* = \{\rho \text{ in } S^1 \mid D^n(\rho) \neq 0 \text{ for any } n\}.$$

a) Show that $D : S^* \to S^*$ and that D is chaotic on S^*.

b) Let $N(x) = \dfrac{x^2 - 1}{2x}$ be Newton's method for $f(x) = x^2 + 1$ and let

$$\mathbb{R}^* = \{x \text{ in } \mathbb{R} \mid N^n(x) \text{ is defined for all } n\}.$$

That is, $\mathbb{R}^*$ is the set of real numbers that are never mapped to 0 by N. Show that $\phi : S^* \to \mathbb{R}^*$ defined by $\phi(x) = \cot(\frac{x}{2})$ is a homeomorphism.

Hint: It may help to use the previous exercise and the discussion in Example 12.1.

c) Show that $D : S^* \to S^*$ and $N : \mathbb{R}^* \to \mathbb{R}^*$ are topologically conjugate. Conclude that N is chaotic on $\mathbb{R}^*$.

d) Use part (c) to show that N is chaotic on $\mathbb{R}$.

12.17 Show that Theorem 12.15 implies that the number of correct digits to the right of the decimal place approximately doubles with each iteration of Newton's method.

12.18 a) Prove Proposition 12.6 by revising the argument used in Example 12.5.

b) Prove Proposition 12.6 by showing that Newton's function for $f(x) = x^2 - 1$ is topologically conjugate to Newton's function for $q(x) = x^2 - c^2$.

Hint: Use a linear function that maps the fixed points of N_f to the fixed points of N_g as the topological conjugacy.

12.19 Prove Proposition 12.8 by demonstrating that Newton's function for $f(x) = x^2 + 1$ is topologically conjugate to Newton's function for $q(x) = x^2 + c^2$.

12.20 Prove that Newton's methods for $f(x) = ax^3 + bx^2 + cx + d$ and $p(x) = x^3 + \frac{b}{a}x^2 + \frac{c}{a}x + \frac{d}{a}$ are topologically conjugate. Explain why we should not be surprised by this.

12.21 Let $P(x)$ be a polynomial and $Q(x) = P(x + a)$ where a is a real number. Explain why we should expect N_P and N_Q to be topologically conjugate *without* explicitly calculating N_P and N_Q. You may find it useful to draw representative graphs of P and Q. What should the conjugacy be?

12.22 AN OPEN-ENDED PROBLEM: Let $q(x) = ax^4 + bx^3 + cx^2 + dx + e$. Find a fourth-degree polynomial with fewer parameters for which Newton's method is conjugate to N_q. What is the fewest number of parameters necessary? Let $p(x) = (x^2 + A)(x^2 + B)$. For which values of a, b, c, d, and e can we find A and B so that N_q and N_p are topologically conjugate? Investigate the behavior of N_p. What types of behavior do you find?

13
Numerical Solutions of Differential Equations

The reader is probably familiar with the differential equation $y' = ky$ from calculus. Often it is introduced as a simple model for population growth or accumulating interest. For example, if $p(x)$ represents the population of bacteria in a laboratory sample at time x and the population increases by fixed percentage during each time interval, then it is reasonable to assume that there is $k > 0$ such that $p' = kp$; that is, the rate of change of the size of the population, as reflected by the derivative, is proportional to the population. The larger the population, the more individuals that are added to it during each time interval. If we wish a more sophisticated model that takes into account the fact that the crowding of a large population may affect growth negatively, then we might reflect this by adding a quadratic term to get $w' = kw - aw^2$, where both k and a are positive. When the population is small, kw is greater than aw^2, the derivative is positive, and the population is growing. As w increases, the difference between kw and aw^2 decreases until we reach a point where aw^2 is larger than kw, the derivative is negative, and the population is shrinking.

Both of these equations are solvable by techniques we learned in calculus. Suppose that $\frac{dp}{dx} = kp$ and $p(0) = p_0$. Then, by separating variables and integrating we have

$$\int \tfrac{1}{p}\, dp = \int k\, dx \quad \text{or} \quad \log(p) = kx + c.$$

We solve for p and and use the initial condition $p(0) = p_0$ to solve for C to find that

$$p = Ce^{kx} \quad \text{or} \quad p = p_0 e^{kx}. \tag{13.1}$$

Similarly, assume that $w' = kw - aw^2$, $w(0) \geq 0$, and $w(0) \neq \frac{k}{a}$. We separate the variables and integrate $\int \frac{1}{kw^2 - aw} \, dw$ to show that

$$w(x) = \frac{kC}{aC + e^{-kx}} \tag{13.2}$$

is a solution of $w' = kw - aw^2$ where $C = \frac{w(0)}{k - aw(0)}$. If $w(0) = k/a$, then the constant function $p(x) = k/a$ is a solution.

However, not all differential equations are as easy to solve as the two just discussed. It is not hard to create differential equations that are difficult or even impossible to solve. This is not surprising, given the difficulties involved with integrating functions and the fact that not all differential equations can be solved with integration alone. A variety of techniques have been developed to estimate the solutions of differential equations that cannot be solved by other means. The easiest of these is Euler's method, which we illustrate with the following example.

EXAMPLE 13.1.
Let $p' = .5p$ and suppose $p(0) = 1$. Without solving the differential equation, we estimate the value of $p(5)$.

If there is a solution to the differential equation, then we know the slope of the tangent line at each point on its graph. For example, we know that $(0,1)$ is on the graph because $p(0) = 1$, and we know that the slope of the tangent line at $(0,1)$ is $.5(p(0)) = .5$. Also, since the function is differentiable, its graph is close to the tangent line for a small interval around 0. The y-coordinate of the point on the tangent line with x-coordinate 1 can be found by adding the product of the slope and the change in x to $p(0)$. Doing so, we find that when $x = 1$, the y-coordinate on the tangent line is $.5(1) + 1 = 1.5$. Thus, we estimate that $p(1)$ is 1.5. We illustrate this in Figure 13.1.

Now the slope of the tangent to the solution at $(1, p(1))$ is $.5(p(1))$ or approximately $.5(1.5) = .75$. Using the same procedure as before, we obtain a new x value by adding 1 to the current x value, and we estimate a new y value by adding the product of the slope and the change in x to the current y value. That is, we add $(.75)(1) = .75$ to 1.5 to find that $p(2) \approx 2.25$. This is also illustrated in Figure 13.1.

Iterating the procedure, we estimate the slope of the solution at $(2, p(2))$ as $.5(2.25) = 1.125$ and add $(1.125)(1) = 1.125$ to 2.25 to find $p(3) \approx 3.375$. Continuing in this way, we find that $p(5) \approx 7.59375$. A diagram of the

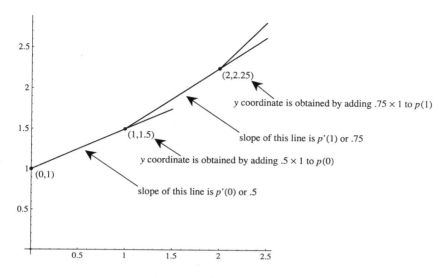

FIGURE 13.1. An illustration of the derivation of the first few points in Euler's approximation of $p' = .5p$ with initial condition $p(0) = 1$.

completed process is shown in Figure 13.2. The reader is encouraged to verify the last two estimates before reading on.

In Figure 13.2, we see that we missed the actual value of $p(5)$ by quite a wide margin. In fact, $p(5) = e^{2.5} \approx 12.18$, and our estimate is about 7.59. It should be apparent that part of the problem is the size of the jumps in x value. In most cases, it is not a good idea to assume that the graph of the solution is more or less the same as the tangent line over an interval of length 1. If we add only .1 to the x value each time, then we must make 50 steps to reach 5 from 0, but we find that our estimate is $p(5) \approx 11.47$, which is closer. By decreasing the size of the steps again to .01, we increase the number of steps to 500, but we find the estimate is now $p(5) \approx 12.17$, a much better approximation. □

Using Example 13.1 as a reference, we describe Euler's method for approximating the solution of the differential equation $p' = f(x, p)$, where x is the independent variable and $f : \mathbb{R}^2 \to \mathbb{R}$. Note that p is a function of x and $f(x, p(x))$ is the slope of the line tangent to the solution passing through the point $(x, p(x))$. Given the initial point $(x_0, p(x_0))$, we develop an algorithm to estimate the value of p at some point $\hat{x}$. For the sake of convenience, we assume $\hat{x} > x_0$, though the method works when $\hat{x} < x_0$. We divide the interval $[x_0, \hat{x}]$ into n subintervals each of length $h = \frac{\hat{x} - x_0}{n}$ and designate the

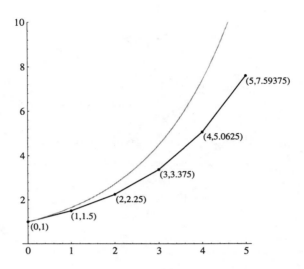

FIGURE 13.2. Estimating a solution of $p' = .5p$ with initial condition $p(0) = 1$ using Euler's method with step size 1. The actual solution is shown in gray for comparison.

endpoints of the subintervals in ascending order by $x_0, x_1, x_2, \ldots x_n = \hat{x}$. Note that $x_{i+1} = x_i + h$ for all i between 0 and n. The number h is called the *step size*, and moving from x_k to x_{k+1} in the approximation process is called a *step*. Of course, it takes n steps to reach $\hat{x}$ from x_0.

Given the value of $p(x_0)$, we estimate the value of $p(x_1)$ by multiplying the slope of the tangent line at x_0 times h and adding the product to $p(x_0)$. Hence, $p(x_1)$ is approximately $f(x_0, p(x_0)) \cdot h + p(x_0)$. This is the same procedure we used in Example 13.1. We repeat the process to approximate $p(x_2)$, arriving at $p(x_2) \approx f(x_1, p(x_1)) \cdot h + p(x_1)$. In general, we have

$$p(x_k) \approx f(x_{k-1}, p(x_{k-1})) \cdot h + p(x_{k-1})$$

or, letting p_k be the approximation of $p(x_k)$,

$$p_k = f(x_{k-1}, p_{k-1}) \cdot h + p_{k-1}. \tag{13.3}$$

We continue with this process until we have reached $p_n \approx p(\hat{x})$. The first few steps in the process are illustrated in Figure 13.3.

If we wish to apply the methods of the previous chapters, we need to find a function g such that $p_n = g^n(p_0)$. Suppose that f is a function of p alone, for example, $f(x,p) = .5p$, as in Example 13.1. Then setting

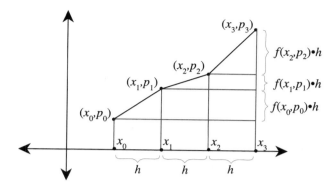

FIGURE 13.3. An illustration of Euler's method.

$g(p) = f(p) \cdot h + p$, we can write equation (13.3) as

$$p_k = g(p_{k-1}) = g(g(p_{k-2})) = \cdots = g^n(p_0).$$

Thus, in the case where f depends only on p, Euler's method is the iteration of a function of the real numbers. If f depends on both x and p, then Euler's method can be modeled as the iteration of a function from $\mathbb{R}^2$ to $\mathbb{R}^2$. Unfortunately, we haven't the tools necessary to completely analyze iterated functions of more than one variable. However, we will see in Example 13.5 that even in the latter case we can still get some information from our understanding of iterated maps of the real line.

EXAMPLE 13.2.

Suppose, as in Example 13.1, that $p' = .5p$, $p(0) = p_0$, and we wish to approximate $p(\hat{x}) = p_n$ by using n steps of size $h = \frac{1}{n}(\hat{x})$. Then

$$p_n = (1 + .5h)^n p_0 \text{ or } p_n = g^n(p_0),$$

where $g(p) = (1 + .5h)p$.

To see this, we note that from equation (13.3),

$$p_k = f(p_{k-1}) \cdot h + p_{k-1} = .5p_{k-1}h + p_{k-1} = (1 + .5h)p_{k-1}.$$

So, if we let $g(p) = (1 + .5h)p$, then

$$\begin{aligned}
p_n &= (1 + .5h)p_{n-1} &&= g(p_{n-1}) \\
&= g(1 + .5h)p_{n-2} = g^2(p_{n-2}) \\
&\vdots \\
&= (1 + .5h)^n p_0 &&= g^n(p_0).
\end{aligned}$$

Recall from calculus that $\lim_{n \to \infty} \left(1 + \frac{r}{n}\right)^n = e^r$. So, since $h = \frac{\hat{x}}{n}$,

$$\lim_{h \to 0} g^n(p_0) = \lim_{h \to 0} (1 + .5h)^n p_0 = \lim_{n \to \infty} \left(1 + .5\frac{\hat{x}}{n}\right)^n p_0 = e^{.5\hat{x}} p_0.$$

Therefore, the estimate derived from Euler's method converges to the solution as the step size approaches 0, and the number of steps goes to infinity. □

We demonstrated in Example 13.2 that the Euler approximation of the value of the solution of $p' = kp$ at any point converges to the value of the actual solution at that point, as the step size converged to zero and the number of steps went to infinity. Usually, we see a similar behavior: the Euler approximation converges to the actual value at a fixed point in the domain as step size decreases. However, there are limits beyond which smaller step sizes are no longer practical because approximations with large numbers of steps must be mechanized, and computers have finite accuracy. To a computer, there are only finitely many numbers between 0 and 1. This limitation introduces possibly fatal errors as the step size becomes smaller. The interested reader is referred to the text *Differential Equations: A Dynamical Systems Approach* by J. Hubbard and B. West for details.

We might also ask whether the Euler approximation for a fixed step size has the same qualitative behavior as the solution of the differential equation. By qualitative behavior, we mean the general shape of the solution curve. For example, if the solution is $p(x) = Ce^{-x}$, then we know that all solutions tend to 0 as x tends to infinity. Or if the solution is $p(x) = \sqrt{x} + \frac{C}{x}$, then we know that the solution tends to $\sqrt{x}$ as x tends to infinity. This kind of information can sometimes be more useful than knowledge of the value of the solution at a specific point. For example, suppose we know that the population of a given mammal in an ecosystem is modeled by a differential equation whose general solutions have the form $p(x) = 150\sin(\frac{\pi}{4}x) + \frac{450x+C}{x}$, where x is in years and $p(x)$ is the number of individuals in year x. Then we won't be unduly alarmed if the population drops to 325 individuals from a high of 575 in a three-year time span since we know that the qualitative behavior of the solutions is to oscillate between highs of 600 and and lows of 300 with a period of eight years.

In the remainder of this chapter, we will focus on the qualitative behavior of Euler approximations.

EXAMPLE 13.3.
Consider $p' = kp$ where k is any real number and $p(0) = p_0$. As we have seen, the actual solution is $p(x) = p_0 e^{kx}$. Suppose that we approximate the solution by using Euler's method with fixed step size $h > 0$. We

13. Numerical Solutions of Differential Equations   159

denote successive approximations, $p_0 = p(0)$, $p_1 \approx p(h)$, $p_2 \approx p(2h)$, ... An analysis similar to the one used in the preceding example, demonstrates that $p_n = kp_{n-1}h + p_{n-1}$. That is, $p_n = (1 + kh)^n p_0$ or $p_n = g^n(p_0)$, where $g(p) = (1 + kh)p$.

We note that g is a linear function in p, the only fixed point of g is 0, and that $|g'(p)| = |1 + kh|$. If $k > 0$, then $|g'(p)| > 1$ for all p. Thus, $g^n(p_0)$ tends to infinity whenever $p_0 \neq 0$ and remains constant at 0 when $p_0 = 0$. Qualitatively, this is the same behavior we see in the solutions of the differential equation; when $k > 0$, the solution $p(x) = p_0 e^{kx}$ tends to infinity as x increases and is identically zero when $p_0 = 0$.

Now consider $k < 0$. If $0 < h < -\frac{2}{k}$, then $g'(p) = |1 + kh| < 1$. Hence, if h is in this range, then $g^n(p_0)$ tends to 0 for all p_0 and $W^s(0)$ is the set of all real numbers. If $h = -\frac{2}{k}$, then $|g'(p)| = 1$ and $g^n(p_0) = p_0$ for all n. Finally, if $h > -\frac{2}{k}$, then $g^n(p_0)$ tends to infinity. Which behavior is correct? Checking the solution we see that when $k < 0$, $p(x) = p_0 e^{kx}$ tends to 0 as x goes to infinity so Euler's approximation has the same qualitative behavior as the solution if and only if $h < -\frac{2}{k}$. How small the step size h needs to be in order to guarantee the correct qualitative behavior depends on the absolute value of k; no single h works for all k. The larger k is in absolute value, the *smaller* h must be to exhibit the correct behavior. □

In Examples 13.2 and 13.3, Euler's approximation reduced to iteration of a linear function. If we investigate more complicated differential equations, we expect to see more complicated dynamics, and indeed this is what happens.

EXAMPLE 13.4.
Let $p' = p - p^2$, $p(0) = p_0$, and let h represent a positive step size for a Euler approximation. Again, we denote the estimated value of $p(nh)$ by p_n. If $g(p) = (p - p^2)h + p$, then

$$p_n = (p_{n-1} - p_{n-1}^2) h + p_{n-1} = g(p_{n-1}) = g^n(p_0).$$

In Exercise 9.5, we saw that all quadratic maps are topologically conjugate to maps of the form $q_c(p) = p^2 + c$ via a linear conjugacy. In particular, there exists $\tau(p) = mp + b$ such that $\tau \circ q_c = g \circ \tau$. A few computations reveal that $g(p) = (p-p^2)h+p$ is conjugate to $q_c(p) = p^2 + c$ where $c = \frac{1-h^2}{4}$. Since $h > 0$, there is a positive value of h for which the qualitative behavior of $g(p)$ is the same as the qualitative behavior of $q_c(p) = p^2 + c$ for each $c < \frac{1}{4}$. Recall from Exercise 10.7 that when $-3/4 < c < 1/4$, the only periodic points of q are a single attracting fixed point and a repelling fixed point. However, when $c < -3/4$, we see a cascade of period-doubling bifurcations followed by chaos. Consequently, the qualitative behavior of the Euler approximation depends on the choice of h and hence c. We now ask how this corresponds to the qualitative behavior of the solutions of the differential equation.

From equation (13.2), we know that the solution of $p' = p - p^2$ has the form $p(x) = \frac{C}{C+e^{-x}}$ where $C = \frac{p(0)}{1-p(0)}$. Therefore, if $p(0)$ is not 1 or 0, then $p(x)$ tends to 1 as x tends to infinity. If $p(0) = 1$, then the constant function $p(x) = 1$ is a solution that attracts all other solutions, and if $p(0) = 0$, then the solution is the constant function $p = 0$ that repels all other solutions. Returning to the Euler approximation, we are led to ask if the solutions $p = 1$ and $p = 0$ correspond to the attracting and repelling fixed points we see when $-3/4 < c < 1/4$.

We note that $\frac{1-\sqrt{4c}}{2}$ is an attracting fixed point of q_c and $\frac{1+\sqrt{4c}}{2}$ is a repelling fixed point when $-3/4 < c < 1/4$. In the exercises, we will ask the reader to show that the topological conjugacy maps $\frac{1-\sqrt{4c}}{2}$ to 1 and $\frac{1+\sqrt{4c}}{2}$ to 0. Therefore, 1 is an attracting fixed point of Euler's approximation and 0 is a repelling fixed point when $-3/4 < c < 1/4$. Hence, the qualitative behavior of the approximation is the same as the qualitative behavior of the solution when c is in this range. Substituting $c = \frac{1-h^2}{4}$ into $-3/4 < c < 1/4$ and simplifying, we get $0 < h < 2$, and we know that Euler's approximation displays the correct behavior when $h < 2$. If $h > 2$, then $c < -\frac{3}{4}$ and the approximation still has fixed points at 0 and 1, but they are both repelling. As h tends towards infinity, c tends towards negative infinity and the approximation goes through a series of period-doubling bifurcations. When $h > 3$, $c < -2$ and, by Exercise 10.5, the Euler approximation is chaotic on a Cantor set with the remaining orbits tending to infinity. A few typical behaviors are illustrated in Figure 13.4.

As a concluding comment, we note that if $p_0 = 3$ and $h = 1$, then the Euler approximation does not converge to 1, even though the value of h is within the range for which 1 is an attracting fixed point. We will explore this apparent contradiction in Exercise 13.2. □

In the previous examples, the derivative of the solution of the differential equation at any point depended only on the value of the solution. We saw that for a small enough value of h the Euler approximation always exhibited the correct qualitative behavior. In the following example we will see that it is not always possible to find a value of h small enough to guarantee the correct qualitative behavior of the approximation.

EXAMPLE 13.5.

Now consider the differential equation $p' = -xp$. By using techniques similar to those we used earlier, we find that the solution of this equation

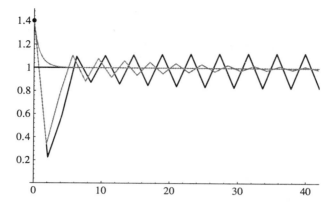

FIGURE 13.4. The light curve is the graph of the solution of $p' = p - p^2$ with initial condition $p(0) = 1.4$. The lighter jagged line is the Euler approximation for $h = 1.9$, and the dark jagged line is the Euler approximation for $h = 2.1$. Note that the approximation when $h = 2.1$ appears to converge to a period two cycle whereas the approximation when $h = 1.9$ tends to 1, as does the actual solution.

is $p(x) = Ce^{-x^2/2}$ where $C = p(0)$. However, computing the Euler approximation in this case is a bit more complicated since the value of the derivative is dependent on both the value of x and the value of p. We extend our notation to include this fact by labeling the initial point in the approximation $(0, p_0)$. Then the first iterate of Euler's approximation with step size h yields (h, p_1), the second yields $(2h, p_2)$, and so forth, with the nth iterate being (nh, p_n). Note that as before we increase the first coordinate by the step size h with each iterate. Also, the slope of the tangent line at the point (nh, p_n) is $-(nh)p_n$. To determine p_{n+1}, we use equation (13.3). We have

$$p_{n+1} = -((nh)p_n)h + p_n = (1 - nh^2)p_n.$$

To write the Euler approximation with step size h as the iteration of a function, as we have in the previous two examples, we must employ the function $G : \mathbb{R}^2 \to \mathbb{R}^2$ defined by

$$G(x, p) = (x + h, p - xph).$$

Then $(nh, p_n) = G^n(0, p_0)$. Even though G is a function of $\mathbb{R}^2$, we will use the tools we have developed for investigating iterated functions of $\mathbb{R}$ to determine whether or not the Euler approximation exhibits the correct qualitative behavior.

We note that the actual solution, $p(x) = Ce^{-x^2/2}$, tends to 0 as x tends to infinity regardless of the initial value chosen. Setting the step size at .2 and the initial value at $p(0) = 1$, we use a computer to test the asymptotic behavior of Euler's method for the first 200 iterates or so. We find

$$p_0 = 1 \qquad\qquad p_{60} \approx -10^{-35}$$
$$p_{10} \approx 10^{-1} \qquad\qquad p_{70} \approx -10^{-33}$$
$$p_{20} \approx 10^{-5} \qquad\qquad p_{80} \approx -10^{-30}$$
$$p_{30} \approx -10^{-30} \qquad\qquad p_{90} \approx 10^{-26}$$
$$p_{40} \approx -10^{-34} \qquad\qquad p_{100} \approx 10^{-22}$$
$$p_{50} \approx -10^{-35}$$

This is more or less the behavior we expected, though the rise in the value of p for the last few iterations is worrisome. Looking ahead, we see things aren't getting any better; in fact they are substantially worse:

$$p_{120} \approx -10^{-11} \quad \text{and} \quad p_{140} \approx -10^2$$

Finally, we have

$$p_{197} \approx 10^{44} \qquad p_{198} \approx -10^{45} \qquad p_{199} \approx 10^{46} \qquad p_{200} \approx -10^{47}$$

Clearly, things are getting out of hand. The iterations are swinging wildly from one side of 0 to the other, and the magnitude of the swings is growing. Perhaps it will settle down again, but it doesn't look good. Taking a hint from our earlier examples, we try to avoid this unpleasant behavior by using a smaller step size. In the past, a small enough step size was all that we needed to guarantee that the qualitative behavior of the Euler approximation was the same as that of the actual solution. Letting $h = .1$ we try again. We find

$$p_{100} \approx 10^{-43}$$
$$p_{200} \approx -10^{-99}$$
$$p_{300} \approx -10^{-83}$$
$$p_{500} \approx -10^{11}$$
$$p_{1000} \approx -10^{412}.$$

Things are definitely not getting better.

Based on the limited evidence we have acquired so far, one might speculate that over the long term, Euler's approximation does not behave in the same way as the actual solution. Closer examination of the iterated function bears this out. Recall that $(hn, p_n) = G^n(0, p_0)$ where

$G(x,p) = (x + h, (1 - xh)p)$. If we write the coordinate functions separately, we have $G(x,p) = (g_1(x,p), g_2(x,p))$ where $g_1(x,p) = x + h$ and $g_2(x,p) = (1 - xh)p$. As we are principally concerned with the fate of p_n as n tends to infinity, we focus our attention on the behavior of g_2 under iteration.

If p_0 is an initial value that remains unchanged under iteration of G, then $g_2(x, p_0) = p_0$. Thus, $p_0 = (1 - xh)p_0$ and $(-xh)p_0 = 0$. Since we have assumed h is not 0 and x cannot be 0 for more than one iteration, this implies $p_0 = 0$ is the only initial condition which yields a stable state. This corresponds to the behavior of the solution of the differential equation in which the constant equation $p(x) = 0$ is a solution. However, in the differential equation all other solutions tend to this constant solution over time; the value of the solution, $p(x) = Ce^{-x^2/2}$, to the differential equation $p' = -xp$ tends to 0 as x tends to infinity. On the other hand, for the two values of h we have checked, the Euler approximation tended to 0 for a time and then began to grow.

To determine if the growth of the Euler approximation continues in an unbounded fashion for all values of h, we again consider $g_2(x,p) = (1-xh)p$. Note that as x increases, $(1-xh)$ decreases. If $x > 2/h$, then $(1-xh) < -1$, and since x continues to grow as we iterate G, the behavior of g_2 is similar to that of $\ell(p) = mp$ where $m < -1$: all points but 0 are in the stable set of infinity and oscillate from side to side of zero. Hence, no matter how small we choose h, once $x > 2/h$, that is, once we have completed $2/h^2$ iterations, the Euler approximation begins to grow without bound. We will make this statement precise in Exercise 13.5. Note that this analysis predicts that the approximation will begin to grow after 50 iterations when $h = .2$ and after 200 iterations when $h = .1$. This correlates well with the experimental data shown earlier. □

Exercise Set 13

13.1 a) Complete the steps outlined in Example 13.1 and verify the estimate for $p(5)$.

b) Let $p' = .5p$ and $p(0) = 1$. Using Euler's method with five steps each of length .2, estimate $p(1)$ without using a computer.

13.2 In Example 13.4, we claimed that there is a linear map that is a topological conjugacy between the Euler approximation of the differential equation $p' = p - p^2$ with step size $h > 0$ and the quadratic map $q(x) = x^2 + c$ where $c = \frac{1-h^2}{4}$.

a) Find the topological conjugacy.

b) Show that 0 and 1 are always fixed points of the Euler approximation and that 1 is an attracting fixed point if and only if $0 < h < 2$.

c) If $h = 1$ and $p_0 = 3$, the Euler approximation grows without bound. Why is this?
   When $h = 1$, what is $W^s(1)$?

d) Determine the stable set of 1 for each h between 0 and 2. Given $p_0 > 1$, find a formula that determines for which h the Euler approximation with initial value p_0 will converge to 1.

e) Determine the values of h for which the Euler approximation has an attracting periodic orbit with prime period two. Choose a step size in the range you have found, and use a computer program to demonstrate several initial values that are attracted to a period two orbit. What are the points in the period two orbit?

f) Explain why there exists h_0 such that the Euler approximation has periodic points of all prime periods if and only if $h > h_0$. Estimate h_0 to the best of your ability.

13.3 For each of the following differential equations, represent the Euler approximation as the iteration of a function of the real numbers. Determine for which step sizes the qualitative behavior of the approximation is the same as the qualitative behavior of the solutions.

a) $p' = -\frac{1}{2}p$

b) $p' = p - 2p^2$

c) $p' = -p^2$
   Note: For part (c), if $p(0) \neq 0$, then the solution is $p(x) = \frac{1}{x+c}$ where $c = \frac{1}{p(0)}$ and if $p(0) = 0$, then the solution is the constant function $p = 0$.

13.4 In Example 13.3, we claimed that when $p' = kp$ and $k < 0$, the larger k is in absolute value, the smaller the step size must be in order for the Euler approximation to exhibit the correct qualitative behavior. Use derivatives to explain why this is so. For which values of k must we use a step size smaller than .01 in order to guarantee the correct qualitative behavior?

Exercise Set 13   165

13.5 In Example 13.5, we claimed that if $x > 2/h$, then the growth of the second coordinate of the function $G(x,p) = (x+h, (1-xh)p)$ under iteration of G behaves qualitatively like iteration of the function $\ell(p) = -mp$ where $m > 1$. In particular, as n tends to infinity, the absolute value of the second coordinate of $G^n(x,p)$ also tends to infinity. To make this precise, fix $h > 0$, let (x_0, p_0) be an initial point satisfying $x_0 > 2/h$, and define $\ell(p) = (1-x_0h)p$. Prove that the absolute value of the second coordinate of $G^n(x_0, p_0)$ is greater than or equal to $|\ell^n(p_0)|$ for all natural numbers n.

Hint: Let $g_2(x,p) = (1 - xh)p$ and prove that

$$|g_2(G^n(x_0, p_0))| \geq \ell^{n+1}(p_0) = |1 - x_0h|^{n+1}p_0.$$

13.6 The solution of the differential equation $p' = p - x$ is

$$p(x) = x + 1 + Ce^x \text{ where } C = p(0) - 1.$$

Determine whether or not the qualitative behavior of the Euler approximation is the same as that of the solutions of the equation. If not, why does it go awry?

14

The Dynamics of Complex Functions

Many interesting and beautiful results in dynamics are seen in the realm of complex functions. In the last two chapters of this text, we will examine Newton's method for complex functions and the dynamics of the quadratic map $q_c(z) = z^2 + c$ when z is complex. Both of these subjects yield interesting mathematics and beautiful illustrations. In particular, the well-known Mandelbrot set is intimately connected with the dynamics of the quadratic map.

14.1. The Complex Numbers

The set of complex numbers consists of all numbers of the form $a + bi$ where a and b are real numbers and i satisfies $i^2 = -1$. We denote this set by the symbol $\mathbb{C}$. If $a + bi$ and $c + di$ are two complex numbers, then we define addition, multiplication, subtraction, and division as follows:

(1) $(a + bi) + (c + di) = (a + c) + (b + d)i$
(2) $(a + bi) - (c + di) = (a - c) + (b - d)i$
(3) $(a + bi)(c + di) = ac + bci + adi + bdi^2 = (ac - bd) + (ad + bc)i$
(4) $(a + bi) \div (c + di) = \dfrac{a + bi}{c + di}$
$= \dfrac{(a + bi)(c - di)}{(c + di)(c - di)} = \left(\dfrac{ac + bd}{c^2 - d^2}\right) + \left(\dfrac{bc - ad}{c^2 - d^2}\right)i$

Note the use of a technique reminiscent of "rationalizing the denominator" when simplifying the quotients of complex numbers.

14. The Dynamics of Complex Functions

While the algebraic properties of complex numbers were known as early as the sixteenth century, their importance and utility were not realized until their geometric properties were developed by Gauss in the nineteenth century. We represent the complex number $a + bi$ in the Cartesian coordinate system by the point (a, b). For example, the number $2 + 3i$ corresponds to the point $(2, 3)$ in the plane. The set of real numbers considered as a subset of the complex plane lies on the x-axis, and the set of pure imaginary numbers, which is all complex numbers of the form bi, falls on the y-axis. Consequently, in the complex plane we often call the horizontal axis the real axis and the vertical axis the imaginary axis.

A complex number plotted as a point in the plane can profitably be visualized as a vector extending from the origin to this point. Adding two complex numbers is analogous to adding the corresponding vectors. This interpretation also leads to a natural definition of length. The complex number $a + bi$ has a length, or *modulus*, that is the distance from the origin to the point (a, b) and is computed as $\sqrt{a^2 + b^2}$. This is the same notion of length typically employed for vectors. We denote the modulus of $a + bi$ as $|a + bi|$. It is no accident that the notation used is the same as the absolute value notation in the real numbers. The absolute value of a real number can also be thought of as the distance from that number to zero. We define a metric on the complex numbers in the same way that we did for real numbers. Let z and w be complex numbers. Then the distance from z to w is $|z - w|$. Or, using our earlier notation, if $z = a + bi$ and $w = c + di$, then the distance from z to w is

$$|z - w| = |(a + bi) - (c + di)|$$
$$= |(a - c) + (b - d)i| = \sqrt{(a - c)^2 + (b - d)^2}.$$

We leave the proof that this definition of distance does in fact form a metric on the complex numbers as an exercise.

Now that we have a metric on $\mathbb{C}$, we can describe its topology using Definition 11.4. In particular, we note that the neighborhood

$$N_\epsilon(z) = \{w \text{ in } \mathbb{C} \mid |z - w| < \epsilon\}$$

is a disk in the plane with center z and radius ϵ. Hence, an open set in the complex plane must contain a disk around each point in the set. Readers for whom this is a first introduction to complex analysis are encouraged to review Definition 11.4 and see that they understand it as it applies to the complex plane.

Points in the plane may be represented by polar coordinates as well as the system of rectangular coordinates used in the Cartesian coordinate system. If $a + bi$ is a point in the plane, then its position is completely determined by its distance from the origin and the angle from the positive x-axis to

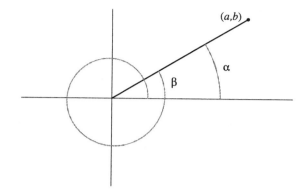

FIGURE 14.1. The complex number (a,b) is shown with its argument indicated. The *argument* is the angle from the positive x-axis to the ray through (a,b). Two such angles are shown here.

the ray extending from the origin through $a + bi$. To be more specific, the location of the point (a, b) is determined by its modulus, $\sqrt{a^2 + b^2}$, and an angle α satisfying

$$\cos(\alpha) = \frac{a}{\sqrt{a^2 + b^2}} \quad \text{and} \quad \sin(\alpha) = \frac{b}{\sqrt{a^2 + b^2}}. \qquad (14.1)$$

This angle is usually called the *argument* of $a+bi$ and is denoted $\arg(a + bi)$. Of course, there are a few problems with the definition of the argument. First, there are infinitely many angles that satisfy equation (14.1). Two of these are demonstrated in Figure 14.1. Second, the argument isn't defined when $(a, b) = (0, 0)$. The first of these problems is easy to deal with; we just need to remember that if two complex numbers have the same modulus and their arguments differ by an integer multiple of 2π, then they are the same number. For example, $\sqrt{3}+i$ has modulus 2 and argument $\frac{\pi}{6} + 2n\pi$, where n can be any integer. We acknowledge the second problem by not defining an argument for zero. (There appear to be several good candidates for the argument of zero, but all of them lead to contradictions in complex function theory. Hence, we must resign ourselves to not being able to express zero in polar coordinates. As we will see, this is not an undue hardship.)

Now let z be a nonzero complex number, r be the modulus of z, and α be the argument of z. Then it should be clear that

$$z = r(\cos(\alpha) + i \sin(\alpha)). \qquad (14.2)$$

If we extend the exponential function to the complex numbers so that it is differentiable at all points in $\mathbb{C}$ and $e^u e^v = e^{u+v}$ for all complex numbers

170   14. The Dynamics of Complex Functions

u and v, then it can be shown that

$$\cos(\alpha) + i\sin(\alpha) = e^{\alpha i}. \tag{14.3}$$

This equation is often called *Euler's formula*. Combining equations (14.2) and (14.3), we arrive at

$$z = re^{\alpha i}$$

where $r = |z|$ and $\alpha = \arg(z)$. We call this representation of z the *exponential* or *polar form*.

When expressed in exponential form, a geometric interpretation of multiplication and division is readily apparent. If $z = re^{\alpha i}$ and $w = se^{\beta i}$, then $zw = (rs)e^{(\alpha+\beta)i}$. On the other hand, $z \div w = \dfrac{re^{\alpha i}}{se^{\beta i}} = \dfrac{r}{s}e^{(\alpha-\beta)i}$. Thus, the modulus of the product of two complex numbers is the product of the moduli of the factors, and the argument of the product is the sum of the arguments of the factors. A similar statement holds for the quotient of complex numbers. We can now find the square roots of complex numbers.

EXAMPLE 14.1.
Find the square roots of $z = 4e^{\frac{\pi}{4}i}$.
If we indicate the root by w, then $|w|^2 = 4$ and $2\arg(w) = \frac{\pi}{4}$. Obviously, $|w| = 2$ and $\arg(w) = \frac{\pi}{8}$ suffice. How do we find the second root?
Recall that we may write $\arg(z) = \frac{\pi}{4} + 2n\pi$ where n is any integer. So the equation we really need to solve is $2\arg(w) = \frac{\pi}{4} + 2n\pi$. Dividing by 2, we arrive at $\arg(w) = \frac{\pi}{8} + n\pi$. Writing this as a real number plus integer powers of 2π, we see that $\arg(w) = \frac{\pi}{8} + 2n\pi$ or $\arg(w) = \frac{9\pi}{8} + 2n\pi$. So the roots of $z = 4e^{\frac{\pi}{4}i}$ are $w_1 = 2e^{\frac{\pi}{8}i}$ and $w_2 = 2e^{\frac{9\pi}{8}i}$. The complex number z and its roots are shown in Figure 14.2. □

14.2. Complex Functions

To aid our investigation of the dynamics of complex functions we restate a few of the definitions and theorems that we developed for functions of a real variable. In what follows, the domain and range of all functions are assumed to be subsets of the set of complex numbers unless stated otherwise.

DEFINITION 14.2. *The function $f : D \to C$ is continuous at the point z_0 in D if for each $\epsilon > 0$ there exists $\delta > 0$ such that $|f(z_0) - f(z)| < \epsilon$ for all z in D satisfying $|z_0 - z| < \delta$. A function is continuous if it is continuous at each point in its domain.*

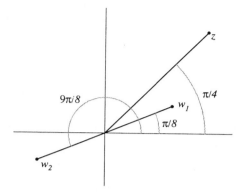

FIGURE 14.2. The geometric representation of the complex number $z = 4e^{\frac{\pi}{4}i}$ and its square roots. Note that if we double the argument of either of the roots we get the argument of z.

Note that this definition is the same as the definition used for functions of the real numbers (Definition 2.7) and functions on metric spaces (Definition 11.4g). In particular, the numbers ϵ and δ are still positive real numbers representing distances. When we visualize continuous functions of the complex plane, we can use the same imagery as when we considered continuous functions of the real numbers. The definition implies that if f is continuous at z_0, then for every neighborhood of $f(z_0)$, there is a $\delta > 0$ such that any point in the neighborhood of z_0 with radius δ is mapped inside the neighborhood of $f(z_0)$. That is, $f(N_\delta(z_0))$ is contained in $N_\epsilon(f(z_0))$. We represent this in the complex plane by drawing the neighborhoods as disks, and we visualize the neighborhood of z_0 being mapped inside the neighborhood of $f(z_0)$. This is illustrated in Figure 14.3.

The definition of the derivative for complex functions is also the same as that of the derivative for real functions.

DEFINITION 14.3. *The function f is differentiable at the point z_0 if f is defined at all points in a neighborhood containing z_0 and the limit*

$$\lim_{z \to z_0} \frac{f(z) - f(z_0)}{z - z_0}$$

exists. In this case, we say f is differentiable at z_0 and call the limit $f'(z_0)$ or the derivative of f at z_0. A function is differentiable if it is differentiable at each point in its domain.

Of course, the problem with this definition is that we must define what

172   14. The Dynamics of Complex Functions

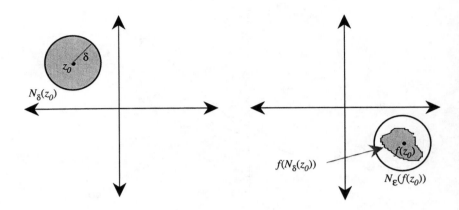

FIGURE 14.3. *The behavior of a function which is continuous at z_0. If f is continuous at z_0, then for each $\epsilon > 0$, there is $\delta > 0$ so that $f(N_\delta(z_0)) \subset N_\epsilon(f(z_0))$.*

we mean by $\lim_{z \to z_0} \frac{f(z)-f(z_0)}{z-z_0}$ for complex valued functions. We recall from calculus that $\lim_{x \to x_0} g(x) = L$ if the value of $g(x)$ is close to L when x is close to a. We make this precise by using the metric.

DEFINITION 14.4. *Let f be a function that is defined on a neighborhood of z_0. Then*

$$\lim_{z \to z_0} f(z) = L$$

if for every $\epsilon > 0$, there exists $\delta > 0$ such that if $0 < |z - z_0| < \delta$, then $|f(z) - L| < \epsilon$.

If δ and ϵ are small, then the expression $0 < |z - z_0| < \delta$ implies that z is close but not equal to z_0 and $|f(z) - L| < \epsilon$ guarantees that $f(z)$ is close to L. Drawing a diagram in the complex plane that illustrates Definition 14.4 can be an instructive exercise, and the reader is encouraged to do so. Notice that we allow z to approach z_0 from any direction. This is a significant difference from the real case in which we can approach numbers only from the left and the right. As a result of this difference, the properties of differentiable complex functions are significantly different from those of differentiable real functions. All differentiable complex functions are *analytic* which means that their Taylor series expansion is defined and converges on a neighborhood of any point in the domain. However, the properties of complex polynomials and rational functions are analogous to those of real polynomials and rational functions. As these are the only types of complex functions we will encounter in this text, we will not develop the general theory of complex functions. Readers who are interested in learning more about the truly fascinating properties of complex functions are encouraged to consult one of the texts listed in the references.

When we discuss the Riemann sphere in Section 14.4, we will find it useful to have a definition of limits that doesn't explicitly refer to a metric. We now restate Definition 14.4 using neighborhoods. The reader should verify that the two statements of the definition are in fact equivalent.

DEFINITION 14.4. (RESTATED) *Let f be a function that is defined on a neighborhood of z_0. Then*
$$\lim_{z \to z_0} f(z) = L$$
if for every neighborhood N of L, there exists a neighborhood N' of z_0 such that if z is in N' and not equal to z_0, then $f(z)$ is in N.

In the following theorems, we collect a few elementary facts about differentiable functions, all of which are analogous to the real case.

THEOREM 14.5. *If a function is differentiable at z_0, then it is continuous at z_0. Further, differentiable functions are continuous.*

THEOREM 14.6. *If f and g are complex functions that are differentiable at z_0, then the following statements hold:*

a) *$f + g$ is differentiable at z_0 and $(f+g)'(z_0) = f'(z_0) + g'(z_0)$.*

b) *If a is a complex number, then $(af)'(z_0) = af'(z_0)$.*

c) *The product fg is differentiable at z_0 and*
$$(fg)'(z_0) = f(z_0)g'(z_0) + f'(z_0)g(z_0).$$

d) *If $g(z_0) \neq 0$, then $\frac{f}{g}$ is differentiable at z_0 and*
$$\left(\frac{f}{g}\right)'(z_0) = \frac{g(z_0)f'(z_0) - f(z_0)g'(z_0)}{(g(z_0))^2}.$$

THEOREM 14.7. THE CHAIN RULE. *If f is differentiable at z_0 and g is differentiable at $f(z_0)$, then the composition $(g \circ f)$ is differentiable at z_0 and $(g \circ f)'(z_0) = g'(f(z_0)).f'(z_0)$.*

THEOREM 14.8. *The formula*
$$\tfrac{d}{dz} z^n = nz^{n-1}$$
holds for all natural numbers n.

We will not prove any of these theorems, but we do illustrate the proof of Theorem 14.8 in the following example. Note that Theorems 14.6 and 14.8 imply that all complex polynomials and rational functions are differentiable throughout their domains. In addition, the formulas for differentiating these functions are analogous to those for real polynomials and rational functions.

EXAMPLE 14.9.
If $f(z) = z^2$, then $f'(z_0) = 2z_0$ for all z_0 since

$$f'(z_0) = \lim_{z \to z_0} \frac{z^2 - z_0^2}{z - z_0} = \lim_{z \to z_0} (z + z_0) \left(\frac{z - z_0}{z - z_0} \right) = 2z_0.$$

□

14.3. The Dynamics of Complex Functions

We begin our investigation of the dynamics of complex functions with functions of the form $f(z) = az$, where a is a complex number. We leave investigation of the slightly more general function $g(z) = az + b$ as an exercise.

EXAMPLE 14.10.
Let $f(z) = az$, where a is a complex number whose modulus is not 1. We note that 0 is the only fixed point of f and investigate the orbit of z_0 when $z_0 \neq 0$.

Clearly, $f^2(z_0) = f(az_0) = a^2 z_0$, and in general $f^n(z_0) = a^n z_0$. If we write a as $|a|e^{\theta i}$ and recall that $a^n = |a|^n e^{n\theta i}$, then we can write $f^n(z_0)$ as $|a|^n e^{n\theta i} z_0$ or in exponential form

$$f^n(z_0) = |a|^n |z_0| e^{(n\theta + \arg(z_0))i}.$$

Therefore, $|f^n(z_0)| = |a|^n |z_0|$, and $\arg(f^n(z_0)) = \arg(z_0) + n\theta$. If $|a| < 1$, then $|a|^n |z_0|$ converges to 0 as n increases. Thus, $W^s(0) = \mathbb{C}$ when $|a| < 1$. On the other hand, $|a|^n |z_0|$ tends to infinity when $|a| > 1$, and so $W^s(0) = 0$ and $W^s(\infty)$ is all of $\mathbb{C}$ except 0 in this case. The argument of a does not affect the convergence or nonconvergence of $f^n(z_0)$ to zero, but it does affect the orbit. Each time the function is iterated, $\arg(a)$ is added to the argument of the input. Thus, when the argument of a is not zero, the orbit of any nonzero point winds around the fixed point. This is illustrated in Figure 14.4. □

14.3. The Dynamics of Complex Functions   175

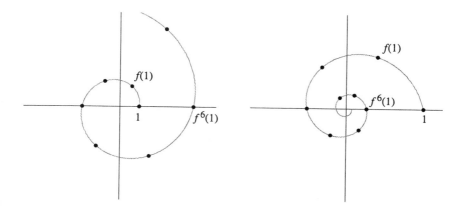

FIGURE 14.4. The orbit of 1 under iteration of the linear function $f(z) = az$. In the first figure, $a = \frac{5}{4}e^{\frac{\pi}{3}i}$, and in the second one $a = \frac{4}{5}e^{\frac{\pi}{3}i}$. Note that in the first diagram the modulus is increased by a factor of 5/4 with each iteration, and in the second diagram the modulus of each iterate is 4/5 of the modulus of the previous iterate. In both cases, $\frac{\pi}{3}$ is added to the argument with each iteration.

In Example 14.10, we saw that 0 is a fixed point of $f(z) = az$ and that $W^s(0) = \mathbb{C}$ when $|f'(0)| = |a| < 1$. On the other hand, $W^s(0) = 0$ when $|f'(0)| = |a| > 1$. This leads us to conjecture that the derivative of a fixed point indicates the repelling or attracting nature of fixed points for complex functions as it does in the real case. We formalize this in the following theorem.

THEOREM 14.11. *Let f be a differentiable complex function and p be a fixed point of f. If $|f'(p)| < 1$, then the stable set of p contains a neighborhood of p. If $|f'(p)| > 1$, then there is a neighborhood of p all of whose points must leave the neighborhood under iteration of f.*

PROOF. We will prove the theorem for the case $|f'(p)| < 1$. The proof of the case $|f'(p)| > 1$ is left as an exercise.

Assume $f(p) = p$ and $|f'(p)| < 1$. Since $\frac{1}{2}(1-|f'(p)|) > 0$, the definitions of the derivative and limits imply that there exists $\delta > 0$ such that if $0 < |z - p| < \delta$, then

$$\left|\frac{f(z) - f(p)}{z - p} - f'(p)\right| < \frac{1}{2}(1 - |f'(p)|). \tag{14.4}$$

Also, by the Triangle Inequality,

$$\left|\frac{f(z) - f(p)}{z - p}\right| \leq \left|\frac{f(z) - f(p)}{z - p} - f'(p)\right| + |f'(p)|. \tag{14.5}$$

Inequalities (14.4) and (14.5) together imply that if $0 < |z - p| < \delta$, then

$$\left|\frac{f(z) - f(p)}{z - p}\right| \leq \frac{1}{2}(1 - |f'(p)|) + |f'(p)| = \frac{1}{2}(1 + |f'(p)|). \tag{14.6}$$

We note that $\frac{1}{2}(1+|f'(p)|) < 1$ since $|f'(p)| < 1$, and set $\lambda = \frac{1}{2}(1 + |f'(p)|)$. Multiplying both sides of inequality (14.6) by $|z - p|$ and substituting λ for $\frac{1}{2}(1 + |f'(p)|)$, we see that when $|z - p| < \delta$, then

$$|f(z) - p| = |f(z) - f(p)| \leq \lambda |z - p|.$$

Continuing by induction, we show that if $|z - p| < \delta$ and n is any natural number, then

$$|f^n(z) - p| \leq \lambda^n |z - p|.$$

Since $\lambda < 1$, this implies that $f^n(z)$ tends to p as n tends to infinity whenever z is such that $|z - p| < \delta$. Thus, the neighborhood of p with radius δ is in the stable set of p. □

Of course, we can generalize Theorem 14.11 to periodic points in the same way that we generalized the analogous theorem for fixed points of the real numbers.

THEOREM 14.12. *Let f be a differentiable complex function and p be a periodic point of f with period k. If $|(f^k)'(p)| < 1$, then there exists a neighborhood of p that is contained in the stable set of p. If $|(f^k)'(p)| > 1$, then there exists a neighborhood of p such that all points must leave the neighborhood under iteration of f^k.*

Theorem 14.12 is an easy extension of Theorem 14.11; its proof is left as an exercise. These theorems motivate the definition of hyperbolic repelling and attracting points.

DEFINITION 14.13. *Let f be a differentiable complex function and p be a periodic point with period k. Then p is a hyperbolic periodic point if $|(f^k)'(p)| \neq 1$. If $|(f^k)'(p)| < 1$, then p is an attracting periodic point. If $|(f^k)'(p)| > 1$, then p is a repelling periodic point.*

As with the real case, nonhyperbolic periodic points do not have predictable behavior in a neighborhood of the point. In the next example, we will examine the behavior of points in in the complex plane when we iterate a function of the form $f(z) = az$, where $|a| = 1$. Note that 0 is the only fixed point of f unless $a = 1$ and 0 is nonhyperbolic since $|f'(0)| = 1$.

14.3. The Dynamics of Complex Functions

EXAMPLE 14.14.

Suppose that $f(z) = e^{\theta i} z$ and z_0 is a nonzero complex number. Then the orbit of z_0 under iteration of f lies on the circle that is centered at the origin and whose radius is $|z_0|$. If θ is a rational multiple of π, then z_0 is a periodic point of f. If the argument of θ is not a rational multiple of π, then z_0 is not a periodic point, and its orbit is dense on the circle.

To see this, note that $f^n(z_0) = e^{n\theta i} z_0$. Consequently,

$$|f^n(z_0)| = |e^{n\theta i}||z_0| = |z_0|$$

for all n, and all iterates of z_0 must be on the circle with radius $|z_0|$ and center at 0. Also, $\arg(f^n(z_0)) = \arg(z_0) + n\theta$ so z_0 is periodic if and only if there exists a natural number n and an integer k such that

$$\arg(z_0) + n\theta = \arg(z_0) + 2k\pi.$$

Solving this equation for θ, we find $\theta = \frac{2k\pi}{n}$. Hence, z_0 is periodic if and only if θ is a rational multiple of π.

Now suppose θ is not a rational multiple of π. We claim that the orbit of z_0 is dense on the circle with radius $|z_0|$ and center at the origin. To prove this, we will show that every neighborhood of every point on the circle with radius $|z_0|$ and center at the origin must contain an iterate of z_0.

Let $\epsilon > 0$ and $r = |z_0|$. Then the circumference of the circle with radius $|z_0|$ and center at the origin is $2\pi r$. As θ is not a rational multiple of π, we know that $f^m(z_0) \neq z_0$ for any natural number m. In particular, if m is larger than $2\pi r/\epsilon$, then none of the first m iterates of z_0 are the same and there must be two between which the distance is less than ϵ. (The reader is encouraged to explain why this is so.)

Suppose that $|f^{m_2}(z_0) - f^{m_1}(z_0)| < \epsilon$ and $m_2 > m_1$. Since we know that $f^{m_1}(z_0) = re^{(m_1\theta + \arg(z_0))i}$ and $f^{m_2}(z_0) = re^{(m_2\theta + \arg(z_0))i}$, this implies that

$$|re^{(m_2\theta + \arg(z_0))i} - re^{(m_1\theta + \arg(z_0))i}| = |e^{(m_1\theta + \arg(z_0))i}(re^{(m_2-m_1)\theta i} - r)|$$
$$= |re^{(m_2-m_1)\theta i} - r| < \epsilon.$$

Thus,

$$|f^{m_2-m_1}(z_0) - z_0| = |re^{(m_2-m_1)\theta i + \arg(z_0)i} - re^{\arg(z_0)i}|$$
$$= |re^{(m_2-m_1)\theta i} - r| < \epsilon. \quad (14.7)$$

As we noted earlier, the function $f^{m_2-m_1}$ rotates points by the nonzero angle $(m_2 - m_1)\theta$ and inequality (14.7) indicates that the distance between successive iterates of $f^{m_2-m_1}$ is less than ϵ. Hence, the sequence of points z_0, $f^{m_2-m_1}(z_0)$, $f^{2(m_2-m_1)}(z_0)$, $f^{3(m_2-m_1)}(z_0)$, ... covers the circle with radius $|z_0|$ by points that are separated by a distance less than ϵ. Thus, if w is any point on the circle of radius $|z_0|$, then the neighborhood of w with radius ϵ must contain an iterate of z_0. As ϵ is arbitrary, this implies that the orbit of z_0 is dense on the circle with its center at the origin and radius $|z_0|$ as desired. □

In the last example, 0 is a fixed point of the function $f(z) = e^{\theta i} z$, $|f'(0)| = |e^{\theta i}| = 1$, and 0 is neutral in the sense that points near 0 do not get closer or further away from 0 as the function is iterated. The reader is encouraged to try to find functions with nonhyperbolic fixed points that either attract or repel nearby points.

We complete this section with an investigation of the dynamics of the function $q(z) = z^2$. We will investigate the more general quadratic family $q_c(z) = z^2 + c$ in Chapter 15.

EXAMPLE 14.15.
Let $q(z) = z^2$. The only fixed points of q are 0 and 1. Since $q'(0) = 0$ and $q'(1) = 2$, 0 is an attracting fixed point and 1 is repelling. If $z = re^{\theta i}$, then $q(z) = (re^{\theta i})(re^{\theta i}) = r^2 e^{2\theta i}$. Iterating, we find $q^2(z) = r^4 e^{4\theta i}$ and in general $q^n(z) = r^{2^n} e^{2^n \theta i}$. Since $|q^n(z)| = r^{2^n}$, we see that z is in $W^s(0)$ if and only if $|z| < 1$ and z is in $W^s(\infty)$ if and only if $|z| > 1$.

If $|z| = 1$, the dynamics are more interesting. Notice that $|z| = 1$ implies that $|q^n(z)| = 1$ for all n. Thus, if we denote the circle of points whose modulus is one by S^1, then we can write $q : S^1 \to S^1$. (The notation S^1 is typically used to denote the unit circle in the plane.) Now if $\arg(z) = \theta$, then $\arg(q(z)) = 2\theta$, and we recognize our friend from Exercise 11.14: the doubling map of the circle. Each time we apply the function $q(z) = z^2$ to a point in S^1, the modulus of the point remains unchanged and the argument is doubled. We demonstrated in Exercise 11.14 that the doubling map on S^1 is chaotic, so it follows that $q(z) = z^2$ is also chaotic on S^1. Orbits of typical points in the complex plane under iteration of q are shown in Figure 14.5. □

14.4. The Riemann Sphere

In this section, we introduce the geometric description of the complex numbers as points on a sphere. While this may seem to be an unnecessary abstraction, we shall see that it is just the notion we need to simplify the proof of several interesting facts in dynamics. For example, in the next section we will use the spherical representation of the complex numbers to show that Newton's method for $f(x) = x^2 + 1$ is chaotic on the real line. While we proved this result already in Example 12.1, we will see that the proof using the Riemann sphere is much simpler. The principal advantage of the spherical representation is that it includes the point at infinity. Once

14.4. The Riemann Sphere

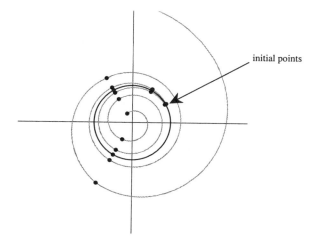

FIGURE 14.5. The orbit of the points $.98e^{\frac{\pi}{6}i}$, $e^{\frac{\pi}{6}i}$, and $1.02e^{\frac{\pi}{6}i}$ under iteration of $q(z) = z^2$. Notice that the orbits converge to zero or infinity if the modulus of the initial point is not equal to one. If the modulus of the initial point is equal to one, then its orbit remains on the unit circle. The unit circle is shown in black, the spirals indicating the orbits are shown in gray.

we have defined this new representation, we will treat ∞ as a point and include it in the domain and range of functions.

To define the Riemann sphere, we use stereographic projection, the same technique that is sometimes used to create maps of the world. A good way to visualize it is to picture a sphere with radius one and center at the origin in $\mathbb{R}^3$. We denote this sphere[1] by S^2. The xy-plane represents the complex plane and intersects the sphere at the equator. This arrangement is pictured in Figure 14.6.

At the top of the sphere is the point $(0, 0, 1)$. If we draw the line through the point at the top and any point in the complex plane, then it intersects the sphere at exactly one point other than $(0, 0, 1)$. If we draw a line from $(0, 0, 1)$ to a point greater than one in modulus, then it intersects the top half of the sphere. If we draw a line through a point which is on the unit

---

[1] We have already used the symbol S^1 to indicate a sphere (circle) with radius one and center at the origin in $\mathbb{R}^2$. The symbol S^2 is used here to indicate a sphere with radius one and center at the origin in $\mathbb{R}^3$. In general, S^n indicates a sphere with radius one and center at the origin in $\mathbb{R}^{n+1}$. Note that position can be defined in terms of one variable (the angle) on S^1; it takes two variables to describe position on S^2, and it takes n variables to describe a position on S^n, hence the superscript.

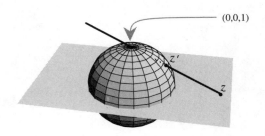

FIGURE 14.6. Stereographic projection of the unit sphere onto the xy-plane. A complex number z and its image z' on the Riemann sphere are shown.

circle, then it intersects at the equator, and if we draw a line from $(0,0,1)$ through a point that is less than one in modulus, it intersects the sphere at a point on the bottom half of the sphere. To get a correspondence between points on the sphere and the points in the complex plane, we assign each complex number to the point on the sphere that is on the line through $(0,0,1)$ and the number. It should be geometrically obvious that this is a one-to-one and onto correspondence between all of the points on the sphere except $(0,0,1)$ and the complex plane. If we also assign $(0,0,1)$ to the point ∞, then we have a one-to-one and onto correspondence between S^2 and $\mathbb{C} \cup \{\infty\}$. We leave as an exercise the verification of the fact that $(0,0,-1)$ corresponds to 0, $(0,1,0)$ corresponds to i, and $(\frac{1}{2}, \frac{1}{2}, \frac{1}{\sqrt{2}})$ corresponds to $(\frac{1}{2}\sqrt{3+2\sqrt{2}})(1+i)$.

Now consider a sequence of points $0, z_1, z_2, z_3, \ldots$ that satisfy $|z_k| = k$. We would like $\lim_{k \to \infty} z_k = \infty$ since it is clear that the points march towards $(0,0,1)$ on S^2. To define this limit, we need to define a set of neighborhoods for each point on S^2. Once we have defined neighborhoods, we can also define open sets, convergence of sequences, and continuous functions without reference to a metric.

DEFINITION 14.16. *Let z' be a point on the Riemann sphere that corresponds to z in the complex plane. The set U' on the Riemann sphere is a neighborhood of z' if the corresponding set U is a neighborhood of z in the complex plane. Let $\epsilon > 0$ and define $N_\epsilon(\infty) = \{z \text{ in } \mathbb{C} \mid |z| > \frac{1}{\epsilon}\}$. Then the corresponding set $N_\epsilon(\infty)'$ on S^2 is a neighborhood of $(0,0,1)$ on the*

Riemann sphere.

Neighborhoods in S^2 correspond directly to the usual definitions of neighborhoods in $\mathbb{C}$ with the addition that neighborhoods of ∞ are defined as sets of the form $N_\epsilon(\infty) = \{z \text{ in } \mathbb{C} \mid |z| > \frac{1}{\epsilon}\}$ where $\epsilon > 0$. It should be clear that as ϵ decreases, the neighborhoods of ∞ also shrink in size. Finally, if we adjoin the point ∞ to $\mathbb{C}$, then it is natural to include ∞ in $N_\epsilon(\infty)$. We caution the reader that it is not wise to select neighborhoods arbitrarily. The foundations of general topology rest on the properties of neighborhoods and open sets, and care was taken when writing Definition 14.16 to see that if N and N' are neighborhoods of x, then $N \cap N'$ is a neighborhood of x, and x is in both N and N'. Readers who are intrigued by the necessity of these seemingly arcane requirements are encouraged to consult a text on general topology. It is a truly fascinating subject.

Since we are defining all neighborhoods in S^2 in terms of sets in $\mathbb{C} \cup \{\infty\}$, we will often refer to the Riemann sphere as the *extended complex plane*. We define functions on the Riemann sphere using the usual algebra of $\mathbb{C}$ with two additional properties. If f is a function defined on a subset of the complex plane, then $f(\infty) = \lim_{z \to \infty} f(z)$ if that limit exists. Also, if $\lim_{z \to a} f(z) = \infty$, then $f(a) = \infty$.

Now that neighborhoods are defined, we restate the definitions of open set, convergent sequence, and continuous function. Note that we are not changing the definitions of these concepts but rather reformulating them without specific mention of a metric. The definitions of accumulation point, closed set, and dense subset are already defined in terms of neighborhoods.

DEFINITION 14.17. *A set is open if every element of the set has a neighborhood that is completely contained in the set.*

DEFINITION 14.18. *A sequence converges to a point if every neighborhood of that point contains the tail of the sequence. That is, the sequence $z_1, z_2, z_3, \ldots$ converges to the point z if for each neighborhood $N_\epsilon(z)$ we can find a natural number M such that $k \geq M$ implies z_k is in $N_\epsilon(z)$.*

DEFINITION 14.19. *Let $f : D \to C$ and z_0 be an element of D. Suppose that for every neighborhood V of $f(z_0)$ we can find a neighborhood U of z_0 such that $f(U) \subset V$. Then f is continuous at z_0.*

Definition 14.17 is a restatement of Proposition 11.5, Definition 14.18 can easily be shown to be equivalent to Definition 11.4c, and Definition 14.19 is clearly equivalent to Definition 14.2. Notice in particular that Figure 14.3 on page 172 can be used to illustrate Definition 14.19 as well as Definition 14.2.

In Section 14.5 and Chapter 15 we will investigate the dynamics of functions of the extended complex plane. While doing so, we will use a special class of functions called *Möbius transformations* or *linear fractional transformations*. These functions have the form

$$T(z) = \frac{az+b}{cz+d}$$

where a, b, c, and d are complex numbers that satisfy the condition that $ad - bc \neq 0$. To extend T to the extended complex plane, we let $T(\infty) = \frac{a}{c}$. Also, if $cz_0 + d = 0$, then $T(z_0) = \infty$. Möbius transformations have many interesting properties. First, they always map circles on the Riemann sphere to circles. Since a line in the complex plane becomes a circle through ∞ in the extended complex plane, this implies that circles and lines in the complex plane are mapped to circles and lines in the complex plane. This is most easily shown by demonstrating that every Möbius transformation can be written as a composition of functions of the form $f(z) = az$, $g(z) = z+b$, and $k(z) = \frac{1}{z}$ and then proving that these functions have the requisite property. Details can be worked out by the reader or found in a text on complex analysis. As an example of this property, we note that the transformation $T(z) = \dfrac{(1+2i)z + 1}{(1-2i)z + 1}$ maps the real axis in the complex plane to the unit circle. The reader may also find it instructive to compute the image of the imaginary axis. The other attribute of Möbius transformations that we will use is that all Möbius transformations are homeomorphisms of the extended complex plane. We leave the proof of this fact as an exercise.

14.5. Newton's Method in the Complex Plane

In Chapter 12, we considered Newton's method on the real line. In this section, we extend that discussion to the complex plane. While the geometric interpretation of Newton's method as an approximation by tangent line is not valid in the complex case, the basic formula is still the same, and the theorems we proved in Chapter 12 still hold. In particular, Newton's function for the complex polynomial $f(z)$ is defined by the equation $N_f(z) = z - \frac{f(z)}{f'(z)}$. Further, we have the following theorem, which is analogous to Theorem 12.2.

THEOREM 14.20. *If $p(x)$ is a complex polynomial and we allow cancellation of common factors in the expression of $N_p(x)$, then $N_p(x)$ is always defined at roots of $p(x)$, a number is a fixed point of $N_p(x)$ if and only if it is a root of the polynomial, and all fixed points of $N_p(x)$ are attracting.*

The proof of Theorem 14.20 is identical to that of Theorem 12.2.

14.5. Newton's Method in the Complex Plane

We begin our investigation of Newton's method for complex quadratic functions with an example.

EXAMPLE 14.21.

Let $f(z) = z^2 - 1$. A quick calculation reveals that $N(z) = \frac{z^2+1}{2z}$. From Theorem 14.20, we know that 1 and -1 are attracting fixed points of N. We claim that if $a + bi$ is an element of $\mathbb{C}$ and $a > 0$, then $a + bi$ is in the stable set of 1. On the other hand, if $a + bi$ is such that $a < 0$, then it is in the stable set of -1. Finally, we claim that N is chaotic on the imaginary axis. We prove this claim by showing that $N(z)$ is topologically conjugate to $q(z) = z^2$ on the extended plane via a Möbius transformation.

If such a Möbius transformation exists, we should be able to find it by searching for one that has the required properties. Since any three points determine a circle and Möbius transformations must map circles to circles, we look for three significant points of N and the corresponding points of q to which they must be mapped. The attracting fixed points of N are 1 and -1, so we map them to 0 and ∞, the attracting fixed points of q. The only other fixed point of N is ∞, so it must be mapped to the repelling fixed point 1 of q. If we are lucky, the coefficient of z in the numerator of the Möbius transformation we are looking for is not zero, and we can divide the numerator and denominator of the transformation by it. In this case, the requisite transformation has the form

$$T(z) = \frac{z+b}{cz+d}.$$

Setting $T(1) = 0$, $T(-1) = \infty$, and $T(\infty) = 1$ and solving for b, c, and d we find that

$$T(z) = \frac{z-1}{z+1}.$$

Verifying that $T \circ N = q \circ T$ is an easy exercise in algebra. As T is a homeomorphism, q and N are topologically conjugate.

Since $T(0) = -1$, $T(i) = i$, and $T(\infty) = 1$, the image of the imaginary axis must be the unit circle. (The unit circle is the only circle passing through -1, i, and 1, and T maps circles and lines to circles and lines.) Since we know from Example 14.15 that q is chaotic on the unit circle and that T is a topological conjugacy, it is tempting to conclude that N is chaotic on the imaginary axis. This is in fact true, but we have to be careful in proving it. Theorem 9.3 states that if a function of a metric space is topologically conjugate to a second function of a metric space that is chaotic, then the first function is also chaotic. But T is a topological conjugacy between functions on the Riemann sphere, which we have not defined as a metric space. As the definition of sensitive dependence on initial conditions uses a metric, the notion of chaotic as we have defined it doesn't even make sense for functions of the Riemann sphere, unless we define a metric there.

We have two alternatives for dealing with this problem. First, we can define the distance between two points on the Riemann sphere by

$$d[(x_1, x_2, x_3), (y_1, y_2, y_3)] = \sqrt{(y_1 - x_1)^2 + (y_2 - x_2)^2 + (y_3 - x_3)^2}.$$

This is the usual Euclidean metric used in $\mathbb{R}^3$. It is easy to show that the subset U of S^2 is open (as defined in Definitions 14.16 and 14.17) if and only if for each x in U there exists a positive real number ϵ such that if y is in S^2 and $d[x, y] < \epsilon$, then y is in U. Thus, we can use this metric to define chaotic behavior on the Riemann Sphere, and Theorem 9.3 holds.

On the other hand, we can avoid these problems by using a topological definition of chaos; we define a continuous function on an infinite set to be *chaotic* if it is topologically transitive on the set and the periodic points are dense in the set. This definition has the advantage of not relying on a metric since the relevant properties are defined only in terms of open sets and are preserved by topological conjugacy. (We should check that these properties are in fact preserved by topological conjugacy when defined in terms of neighborhoods; this is easily done.) Also, Theorem 11.20 implies that for continuous functions on infinite sets, this definition is equivalent to Definition 11.19. Thus, since N on the imaginary axis is topologically conjugate to q on the unit circle, we know that the periodic points of N are dense on the imaginary axis and N is topologically transitive there as well. It follows from our definition that N is chaotic on the imaginary axis.

It remains to show that $a + bi$ is in the stable set of 1 if and only if $a > 0$ and $a + bi$ is in the stable set of -1 if and only if $a < 0$. But $T(1) = 0$ and

$$T(\{a + bi \mid a > 0\}) = \{z \mid |z| < 1\},$$

which is the stable set of 0 under iteration of q. Since T preserves stable sets, the first assertion holds. A similar observation demonstrates the second statement. The reader may wish to verify that the proof that topological conjugacies preserve stable sets does not depend on the existence of a metric in the spaces in question. □

When reflecting on the previous example, we note that the imaginary axis is the perpendicular bisector of the line segment between the roots of $f(z) = z^2 - 1$. Also, recall from our work in Chapter 12 that Newton's method had similar dynamics for large classes of quadratic polynomials. We are led to ask for which complex quadratic polynomials the perpendicular bisector of the line segment between the roots divides the stable sets of the roots and in which of those Newton's method is chaotic on the bisector. In 1879, the English mathematician Arthur Cayley proved that the perpendicular bisector of the line segment joining the roots divides the

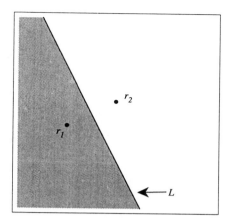

FIGURE 14.7. An illustration of Cayley's Theorem. The roots of the complex quadratic polynomial are indicated by r_1 and r_2. The stable set of r_1 under iteration of $N_q(x)$ is indicated in gray. $N_q(x)$ is chaotic on the perpendicular bisector labeled L. The remaining points are in the stable set of r_2.

stable sets of Newton's function for all complex quadratic polynomials with distinct roots. We can now add that it is also chaotic on the bisector.

THEOREM 14.22. [CAYLEY, 1879] *If the complex quadratic polynomial $q(x)$ has two distinct roots, then $N_q(x)$ is chaotic on the perpendicular bisector of the line segment joining the two roots. Further, the stable set of each root under iteration of $N_q(x)$ is the set of points that lies on the same side of the bisector as the root.*

An illustration of Theorem 14.22 is shown in Figure 14.7. There are two convenient proofs of the theorem. In one, we show that Newton's method for any complex quadratic polynomial with distinct roots is topologically conjugate to Newton's method for $f(z) = z^2 - 1$ via a linear map. In the other, we show that Newton's method for any complex quadratic polynomial with distinct roots is topologically conjugate to the function $f(z) = z^2$ and proceed as in Example 14.21. The details are left to the reader.

We complete the description of Newton's method for complex quadratic polynomials with the following theorem.

THEOREM 14.23. *If the complex quadratic polynomial q has exactly one root, then all points in $\mathbb{C}$ are in the stable set of the root under iteration of $N_q(x)$.*

Theorem 14.23 is easily proven by showing that it holds for $q(z) = z^2$ and then proving that Newton's function for a complex quadratic polynomial with a single repeated root is topologically conjugate to N_q. Theorem 14.23 also generalizes in the obvious way to complex polynomials of higher degree with a single repeated root.

Given our success with understanding Newton's method for complex quadratic polynomials, we are led to hope that an analysis of Newton's method for complex cubics will prove to be as simple. However, our enthusiasm should be tempered by our experience with Example 12.13 in which we found an attracting periodic point with prime period two for Newton's method applied to a real cubic equation. We begin cautiously and investigate Newton's method for $f(z) = z^3 - 1$.

EXAMPLE 14.24.

Let $f(z) = z^3 - 1$. Then $N(z) = \frac{2z^3+1}{3z^2}$ and has attracting fixed points at the roots of $f(z)$, which are 1, $e^{\frac{2\pi}{3}i}$, and $e^{\frac{4\pi}{3}i}$. To get an idea of the dynamics of N, we conduct a computer experiment. We select a grid of points covering the square region whose corners are at $2+2i$, $2-2i$, $-2-2i$, and $-2+2i$. We calculate the value of the 100th iterate of each of these points under iteration of N. If the distance from the 100th iterate to 1 is less than $\frac{1}{4}$, then we assume that the point is in the stable set of 1 and color it black. If the distance from the 100th iterate to $e^{\frac{2\pi}{3}i}$ is less than $\frac{1}{4}$, then we assume that the point is in the stable set of $e^{\frac{2\pi}{3}i}$ and color it gray. All other points are assumed to be in the stable set of $e^{\frac{4\pi}{3}i}$ The result of this experiment using a 250 by 250 grid is shown in Figure 14.8.

We can see that the dynamics of Newton's method for this cubic polynomial are much more complicated than those of Newton's method for quadratic polynomials. In fact, the dynamics of Newton's method for most polynomials of degree higher than two are not completely understood. The reader is encouraged to use experiments similar to the preceding one to try to understand the dynamics of Newton's method for cubics. In particular, we would like to know what happens on the boundary of the stable sets. Specific suggestions for projects can be found in the exercises. □

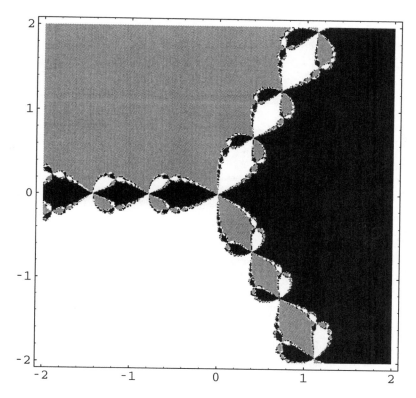

FIGURE 14.8. An analysis of the dynamics of Newton's method for $f(z) = z^3 - 1$. There are three attracting fixed points for this function, 1, $e^{\frac{2\pi}{3}i}$, and $e^{\frac{4\pi}{3}i}$. Points that are known to be in the stable set of 1 are shown in black and points that are known to be in the stable set of $e^{\frac{2\pi}{3}i}$ are shown in gray. All other points are assumed to be in the stable set of $e^{\frac{4\pi}{3}i}$ and are shown in white. The corners of the region shown are at $-2-2i$, $-2+2i$, $2+2i$, and $2-2i$.

Exercise Set 14

14.1 Simplify the following expressions:

a) $(2 - 3i) + (-1 - 4i)$

b) $\frac{1}{2}i - (1 + i)$

c) $(2 + i)(i)$

d) $(2 + i) \div i$

e) $(1 - 4i)(2 + 3i)$

f) $(1 - 4i) \div (2 + 3i)$

14.2 Find the modulus and argument of each of the following complex numbers:

a) $1 - 4i$

b) $\frac{1}{2} + \frac{3}{4}i$

c) 2

14.3 Draw a sketch of the neighborhood of $3 + i$ with radius $\frac{1}{2}$. Do the same for the neighborhood of $-i$ with radius 2.

14.4 Find all of the solutions of the following equations and plot them on the complex plane. What do you notice about the geometry of the solutions?

a) $z^2 = 8$

b) $z^3 - 8 = 0$

c) $z^n - 1 = 0$, for $n = 2, 3, 4, 5,$ and 6

14.5 Let $z_1 = re^{\alpha_1 i}$ and $z_2 = re^{\alpha_2 i}$. Suppose there exists an integer k such that $|r(\alpha_1 + 2k\pi - \alpha_2)| < \epsilon$. Prove $|z_1 - z_2| < \epsilon$.

Hint: Draw a representative diagram and consider the geometry.

14.6 Verify that the function $d[a + bi, c + di] = \sqrt{(c - a)^2 + (d - b)^2}$ is a metric on $\mathbb{C}$.

14.7 Let f be a differentiable complex function. Show that if $f^n(p)$ is defined for all $n \leq k$, then

$$(f^k)'(p) = f'(p) \cdot f'(f(p)) \cdot f'(f^2(p)) \cdot \ldots \cdot f'(f^{k-1}(p)).$$

14.8 Prove that $\dfrac{d}{dz} z^n = nz^{n-1}$ for all integers n.

14.9 Prove that f is continuous at z_0 if it is differentiable there.

14.10 a) Complete the details of the induction argument used in the proof of the first half of Theorem 14.11.

b) Let f be a differentiable complex function and p be a fixed point of f. Prove that if $|f'(p)| > 1$, then there is a neighborhood of p all of whose points must leave the neighborhood under iteration of f.

14.11 Let $g(z) = az + b$ where a and b are complex numbers.

a) Find the fixed point of g if it exists. In what case(s) will there not be a fixed point?

b) Show that if $a \neq 1$, then $g(z) = az + b$ is topologically conjugate to a function of the form $f(z) = cz$.

c) In those cases where g has a fixed point, describe its stable set.

d) In those cases where g does not have a fixed point, describe the dynamics of g.

14.12 Plot the orbit of a typical point under iteration of $g_1(z) = \frac{1}{2}z + 1$. Do the same for $g_2(z) = 2z - 2$.

14.13 The orbit of 1 under iteration of $f(z) = az$ for $a = \frac{5}{4}e^{\frac{\pi}{3}i}$ and $a = \frac{4}{5}e^{\frac{\pi}{3}i}$ is shown in Figure 14.4 on page 175. The spiral is drawn so that if u and v are any two points on the spiral that lie on the same ray extending from the origin, then there exists a natural number n such that either $f^n(u) = v$ or $f^n(v) = u$.

a) Find the equation of the spiral when $a = \frac{5}{4}e^{\frac{\pi}{3}i}$ and show that your equation has the required property. Do the same for $a = \frac{4}{5}e^{\frac{\pi}{3}i}$.

b) For which values of a is it possible to draw such a spiral?

c) Prove that the spiral through the iterates of $q(z) = z^2$ shown in Figure 14.5 cannot have this property.

*14.14 Suppose that p is a fixed point of the differentiable function f and $|f'(p)| = 1$. If there exists a neighborhood of p that is contained in the stable set of p, then p is *weakly attracting*. If there exists a neighborhood U of p such that for each point x in U except p there is a positive integer n such that $f^n(x)$ is not in U, then p is *weakly repelling*. Find an example of a function of the complex numbers with a weakly attracting or a weakly repelling fixed point.

14.15 Let $q(z) = z^2$. Show that 1 is a repelling fixed point of q and that $W^s(1)$ is dense on the unit circle in the complex plane. Why doesn't this contradict Theorem 14.11?

14.16 Let $f(z) = z^3$.

a) Find the fixed points of f and determine whether they are attracting, repelling, or nonhyperbolic.

b) Describe the periodic points of f. Determine whether they are attracting, repelling, or nonhyperbolic.

c) Describe the orbit of all the points in $\mathbb{C}$ to the best of your ability.

14.17 Verify that in the description of the Riemann sphere outlined in Section 14.4, $(0, 0, -1)$ corresponds to 0, $(0, 1, 0)$ corresponds to i, and $(\frac{1}{2}, \frac{1}{2}, \frac{1}{\sqrt{2}})$ corresponds to $(\frac{1}{2}\sqrt{3 + 2\sqrt{2}})(1+i)$.

14.18 Show that Definitions 14.18 and 11.4c are equivalent. To do this, let $x_1, x_2, x_3, \ldots$ be a sequence of elements in a metric space, then show that for every $\epsilon > 0$, there exists an integer N such that $d[x, x_k] < \epsilon$ whenever $k \geq N$ if and only if every neighborhood of x contains an element of the sequence.

14.19 Show that Definitions 14.19 and 14.2 are equivalent.

14.20 Let $T(z) = \dfrac{az+b}{cz+d}$ be a Möbius transformation. In particular, $ad - bc \neq 0$.

a) Show that T is defined and continuous on the extended complex plane.
  Note: $T(\infty) = \frac{a}{c}$ and $T(z_0) = \infty$ if and only if $cz_0 + d = 0$.

b) Show that $T : \mathbb{C} \cup \{\infty\} \to \mathbb{C} \cup \{\infty\}$ is one-to-one and onto.

c) Show that T is a homeomorphism of the extended complex plane.

14.21 Prove Theorem 14.20.

Hint: Mimic the proof of Theorem 12.2.

14.22 Prove that Newton's method for $f(z) = z^2 - 1$ is chaotic on the imaginary axis by completing the details of the argument in Example 14.21.

14.23 Let τ be a topological conjugacy between functions that are defined on the extended complex plane. Show that τ maps stable sets of fixed points to stable sets of fixed points without using a metric on S^2.

14.24 Prove Cayley's Theorem (Theorem 14.22), which states that if the complex quadratic polynomial $q(x)$ has two distinct roots, then $N_q(x)$ is chaotic on the perpendicular bisector of the line segment joining the two roots. Further, the stable set of each root under iteration of $N_q(x)$ is the set of points that lies on the same side of the bisector as the root.

Hint: Use one of the suggestions given after the statement of the theorem in the text.

14.25 a) Prove that if a complex quadratic polynomial has exactly one root, then all points in $\mathbb{C}$ converge to the root under iteration of Newton's function for the polynomial. (This is Theorem 14.23.)

b) Describe the dynamics of Newton's function for $f(z) = z^3$ and show that your description is correct. You should be able to completely describe the dynamics on $\mathbb{C}$.

c) Characterize the dynamics of Newton's function for any cubic polynomial with exactly one root of multiplicity three and show that your characterization is correct.

d) Generalize the statement generated in part (c) to higher-degree polynomials and prove that your generalization is correct.

14.26 AN INVESTIGATION: The polynomial $f(z) = z^3 - z^2$ has exactly two roots: 0 is a root of multiplicity two and 1 is a root of multiplicity one. Describe the dynamics of N_f to the best of your ability. If possible, generalize your results to the set of all polynomials with one root of multiplicity two and one root of multiplicity one.

14.27 AN INVESTIGATION: In Example 12.13, we found that the interval of the real line $[-.03, .03]$ is attracted to a period two point of Newton's function for $f(z) = z^3 - 1.265z + 1$. What other points of $\mathbb{C}$ are attracted to this period two point? Can you prove that there is a disk in $\mathbb{C}$ that is attracted to this period two point? This would imply that the set of points that are attracted to a root of f under iteration of N_f is not dense in $\mathbb{C}$. Why? Why is this important?

Can you find a parameter value c for which Newton's function for $f(z) = z^3 - cz + 1$ has an attracting prime period four point? Can you find an open set of parameter values for which this is true?

**Is there a parameter value c for which Newton's method applied to $f(z) = z^3 - cz + 1$ is chaotic on a set that contains an open subset of $\mathbb{C}$?

Note: It suffices to show that an iterate of N_f is conjugate to $q(z) = z^2 - 2$ for some c. You should also explain why this is sufficient as well. Is there an open set of parameter values for which N_f is chaotic on a subset of $\mathbb{C}$?

15
The Quadratic Family and the Mandelbrot Set

We return once again to the study of the dynamics of quadratic functions. In this chapter, we consider the quadratic family $q_c(z) = z^2 + c$. We demonstrated in Exercise 9.5 that all real quadratic functions are topologically conjugate to a real polynomial of the form $q_c(x) = x^2 + c$ for some c. This fact extends to the complex quadratic polynomials; all complex quadratic polynomials are topologically conjugate to a polynomial of the form $q_c(z) = z^2 + c$. (The reader is asked to show this in Exercise 15.1.) We will take direction for our study of the quadratic family from our previous work with the logistic map $h_r(x) = rx(1-x)$.

Recall that the interesting points in the dynamics of h_r are those points remaining in the interval $[0, 1]$ for all time. Points that leave this interval after a finite number of iterations are in the stable set of infinity. Given this, it seems reasonable to suppose that the interesting points in the dynamics of q_c are those that have bounded orbits. An orbit is *bounded* if there exists a positive real number such that the modulus of every point in the orbit is less than this number. The following proposition shows that the orbits of q_c do in fact divide up into two sets: those that are bounded and those that converge to infinity.

PROPOSITION 15.1. *The orbit of a complex number under iteration of a complex quadratic polynomial is either bounded or the number is in the stable set of infinity.*

PROOF. Since all complex quadratic polynomials are topologically conjugate to a function of the form $q_c(z) = z^2 + c$, it suffices to prove the proposition for this family of functions. (In making this assumption, we are using the fact that if $\phi : \mathbb{C} \to \mathbb{C}$ is a homeomorphism and $K \subset \mathbb{C}$ is bounded, then $\phi(K)$ is bounded. We haven't proven this; interested readers are encouraged to do so. A proof can be found in the analysis text by W. Rudin, which is listed in the references.)

We have already shown that the proposition holds for $q(z) = z^2$. (See Example 14.15.) Let c be a nonzero complex number and $q(z) = z^2 + c$. We begin by showing that if w is a complex number satisfying $|w| > |c|+1$, then w is in the stable set of infinity.

Since $q(w) = w^2 + c$,

$$|q(w)| = |w^2 + c| \geq |w^2| - |c|$$
$$\geq (|c| + 1)^2 - |c| = |c|^2 + |c| + 1.$$

Then

$$|q^2(w)| = |q(q(w))|$$
$$= |(q(w))^2 + c|$$
$$\geq |q(w)|^2 - |c|$$
$$\geq (|c|^2 + |c| + 1)^2 - |c|$$
$$= |c|^4 + 2|c|^3 + 3|c|^2 + |c| + 1$$
$$\geq 3|c|^2 + |c| + 1.$$

Continuing by induction, we show that

$$|q^n(w)| \geq (2^n - 1)|c|^2 + |c| + 1.$$

Since the right-hand side of this inequality tends to infinity as n gets larger, our assertion that w is in the stable set of infinity whenever $|w| > |c| + 1$ must hold. It follows that either the orbit of a complex number under iteration of $q(z) = z^2 + c$ is bounded by $|c| + 1$ or the number is in the stable set of infinity. □

The set of points with bounded orbits is interesting enough to have earned its own name.

DEFINITION 15.2. *The set of points whose orbits are bounded under iteration of $q(z) = z^2 + c$ is called the filled Julia set of q and denoted by K_c. The boundary of a filled Julia set is called the Julia set. (The boundary of a set is the collection of points for which every neighborhood contains an element of the set as well as an element not in the set.)*

Julia sets are named after the French mathematician Gaston Julia, who, along with his countryman Pierre Fatou, did much of the fundamental work on the dynamics of complex functions early in the twentieth century. Their results are all the more remarkable given that they did not have the power of computers to help them visualize the sets they were considering. The filled Julia sets for four parameter values are shown in Figure 15.1.

Fatou and Julia proved many interesting theorems. We stated a special case of one of their results in Theorem 10.1. The general statement is given in the following theorem.

THEOREM 15.3. *If a complex polynomial has an attracting periodic orbit, then there must be a critical point of the polynomial in the stable set of one of the points in the orbit.*

We recall that z_0 is a critical point of the polynomial p if $p'(z_0) = 0$. As the only critical point of $q_c(z) = z^2 + c$ is 0, this implies that q_c can have at most one attracting periodic orbit and that 0 must be attracted to it. A proof of Theorem 15.3 can be found in the second edition of Devaney's text *Introduction to Chaotic Dynamical Systems*, which is listed in the references. The role of the critical point is again highlighted in the following theorem of Fatou and Julia.

THEOREM 15.4. *If 0 is in the filled Julia set of $q_c(z)$, then the filled Julia set is connected. That is, given any two points in the filled Julia set, there is a path contained in the set which connects those two points. On the other hand, if 0 is not in the filled Julia set of q_c, then the set is a Cantor set.*

Recall that in Chapter 8, we defined Cantor sets as subsets of the real numbers that are closed and bounded, contain no intervals, and for which every point is an accumulation point of the set. Obviously, this definition is not sufficient for use in Theorem 15.4, so we extend the definition to subsets of the complex plane by introducing the concept of totally disconnected sets.

DEFINITION 15.5. *A set of complex numbers is a Cantor set if it is closed and bounded, totally disconnected, and if every point is an accumulation point of the set. A set of complex numbers is totally disconnected if there does not exist any path between distinct points in the set that is completely contained in the set.*

Proposition 15.1 and Theorem 15.4 naturally divide the complex numbers into two sets: those parameter values c for which the filled Julia set is connected and the orbit of 0 is bounded under iteration of $q_c(z) = z^2 + c$ and those for which the filled Julia set is totally disconnected and 0 is in

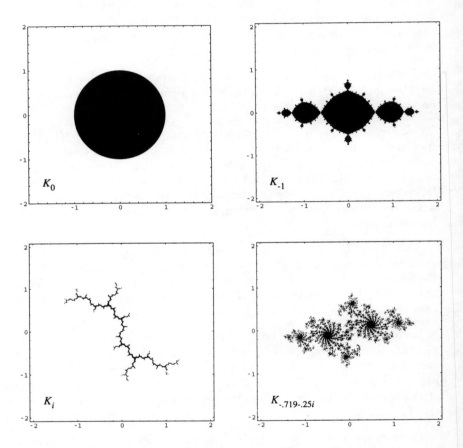

FIGURE 15.1. The filled Julia sets for $c = 0$, $c = -1$, $c = i$, and $c = -.719 - .25i$.

the stable set of infinity. The set of parameter values we will investigate here is the former. More specifically, we are interested in

$$\mathcal{M} = \{c \mid q_c^n(0) \text{ is bounded}\}.$$

While we will not prove Theorem 15.4, in Exercises 15.4 and 15.5 the reader is invited to prove that the filled Julia set is a Cantor set and that $q_c^n(0)$ tends to ∞ when $|c| > 2$. This implies that the set $\mathcal{M}$ is contained inside a disk of radius 2. In Exercise 15.3, we will show that the set of parameter values for which q_c has an attracting fixed point is the interior of the cardioid parametrized by $\rho(\theta) = \frac{1}{2}e^{\theta i} - \frac{1}{4}e^{2\theta i}$. From Theorem 15.3, it follows that this cardioid is contained in $\mathcal{M}$ since 0 must be in the stable set of an attracting fixed point. In fact, if we define

$$\mathcal{M}_k = \{c \mid q_c(z) \text{ has an attracting periodic point with prime period } k\},$$

then

$$\mathcal{M} \supset \bigcup_{k=1}^{\infty} \mathcal{M}_k.$$

In Exercise 15.3, we will show that $\mathcal{M}_2$ is the disk with radius $\frac{1}{4}$ and center -1. It is interesting to try to find the other sets $\mathcal{M}_k$.

So far, we have stated that $\mathcal{M}$ is contained inside the circle of radius 2, $\mathcal{M}$ contains the interior of the cardioid $\rho(\theta) = \frac{1}{2}e^{\theta i} - \frac{1}{4}e^{2\theta i}$, and $\mathcal{M}$ contains the interior of the disk with center -1 and radius $\frac{1}{4}$. To get a better picture of $\mathcal{M}$, we conduct another computer experiment. We know that $\mathcal{M}$ consists of the parameter values for which the orbit of 0 is bounded and that if $|q_c^n(0)| > 2$ for any n, then c is not in $\mathcal{M}$. Thus, we cover the disk of radius 2 by a grid, use each point in the grid as a parameter c, and calculate $q_c^{100}(0)$. If $|q_c^{100}(0)| \leq 2$, we assume that c is in $\mathcal{M}$ and color the corresponding grid point black. If $|q_c^{100}(0)| > 2$, we leave the grid point white. The results of such a program are shown in Figure 15.2.

The reader may recognize the sketch of $\mathcal{M}$ in Figure 15.2 as the famous Mandelbrot set, the first images of which were published by Benoit Mandelbrot in 1980. The Mandelbrot set has been studied intensely since its discovery and much is known about it, though there is still more to be learned. It is known that it is connected, and the manner in which the various bulbs are connected is also understood. The details of this connection are well beyond the scope of this book. An interesting and accessible discussion of the properties of the Mandelbrot set can be found in Chapter 14 of the book *Chaos and Fractals* by Peitgen, Jürgens, and Saupe. It is the author's understanding that a more complete account of the properties of the Mandelbrot set will be included in a preprint on complex dynamics by J. Milnor. Both resources are listed in the references.

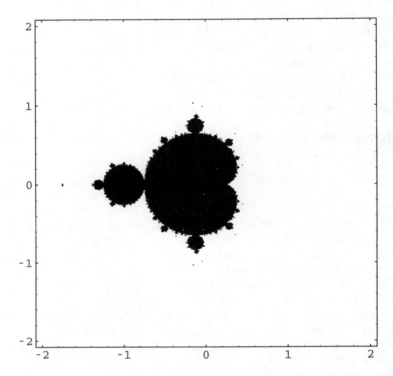

FIGURE 15.2. The set of parameter values c for which $|q_c^{100}(0)| \leq 2$. This set is a good approximation of $\mathcal{M}$. The image contains the cardioid $\rho(\theta) = \frac{1}{2}e^{\theta i} - \frac{1}{4}e2\theta i$ and the disk with radius $\frac{1}{4}$ and center -1 as predicted by our calculations. Note that while M appears not to be connected in this figure, it has been proven that M is connected. This was done by A. Douady and J. H. Hubbard in the mid-1980s.

15.1. Generating Julia and Mandelbrot Sets on a Computer

There are a number of widely available computer programs that generate pictures of Julia sets and the Mandelbrot set. A few of these are listed in the references. Individuals with a minimal amount of programming knowledge will find it easy and instructive to write their own programs. An outline of the *Mathematica* programs the author used to create the illustrations in this text are listed in the Appendix as models of how such programs might be structured. The *Mathematica* programs used by the author are not recommended for creating images of the Julia and Mandelbrot sets as they are not very fast; several of the images created for this text took more than an hour to create in *Mathematica*. Individuals interested in using *Mathematica* for these kind of tasks may wish to consult the text *Applied Mathematica* by W. Shaw and J. Tigg, which is listed in the references. In particular, they demonstrate how the MathLink protocol can be used to reduce the computational time by at least an order of magnitude. As indicated by the preceding statements, the most important factor to consider in writing a program is speed. We outline some of the considerations for efficiently computing the Mandelbrot set here and note that there are similar considerations for Julia sets.

If we wish to create a picture of the Mandelbrot set with a resolution of 256 by 256 and to do so by determining if the 50th iterate of q_c is greater than 2, then we would be foolish indeed to do so by iterating each of the 65,536 points in the grid 50 times. That would involve more than 3.2 *million* iterations. It is quicker to check after each iteration to see if the modulus of that iterate is larger than 2. If it is, then the point is not in the Mandelbrot set and we can stop. We can use this extra check to increase the beauty of our picture by coloring points that leave before five iterations one color, points that leave after five but before ten iterations another color, and so on. A further increase in speed can be gained by automatically including any point that lies inside the cardioid $z = \frac{1}{2}e^{\theta i} - \frac{1}{4}e^{2\theta i}$ or the disk with radius $\frac{1}{4}$ and center -1. Finally, there is no substitute for clever programming. For example, if we wish to know whether $|a + bi| < 2$, then it is computationally more efficient to check whether $a^2 + b^2 < 4$ rather than $\sqrt{a^2 + b^2} < 2$. With these long computations, one is well advised to think before writing.

Finally, we note that there are many algorithms for computing Mandelbrot and Julia sets other than those described here. Several of the alternatives are discussed in the text by Peitgen, Jürgens, and Saupe, which is listed in the references.

Happy programming, and enjoy the exploration of complex analytic dynamics!

Exercise Set 15

15.1 Show that all complex quadratic polynomials are topologically conjugate to a polynomial of the form $q_c(z) = z^2 + c$.
Hint: See Exercise 9.5.

15.2 Complete the details of the induction argument in the proof of Proposition 15.1.

15.3 a) Show that the set of parameter values c for which $q_c(z) = z^2 + c$ has an attracting fixed point is the interior of the cardioid

$$\rho(\theta) = \frac{1}{2}e^{\theta i} - \frac{1}{4}e^{2\theta i}.$$

Hint: We wish to find c so that there exists $z_0 = re^{\theta i}$ such that $z_0^2 + c = z_0$ and $|q_c'(z_0)| < 1$. Substitute for z_0 and solve for c.

b) Show that the set of parameter values for which $q_c(z)$ has an attracting periodic point with prime period two is the interior of the disk with radius $\frac{1}{4}$ and center at -1.

**c) Describe the set of parameter values for which $q_c(z) = z^2 + c$ has an attracting periodic point with prime period k for as many other k as you can. Be forewarned that this is very difficult, and perhaps impossible, to do analytically. Use a computer and experiment. Remember that we have already studied the dynamics of the real function $q_c(x) = x^2 + c$ (in Exercise 10.7) and that those real parameter values for which we have attracting period k points of the real function are also parameter values for the complex function. Good luck!

Note: An answer to this question can be found in Chapter 14 of the book by Peitgen, Jürgens, and Saupe, which is listed in the references.

The following two exercises demonstrate that if $|c| > 2$, then 0 is not in the filled Julia set and the filled Julia set is a Cantor set. This lends support to the statement of Theorem 15.4.

15.4 Let $q_c(z) = z^2 + c$.

a) Prove that if $|z| \geq |c|$ and $|z| > 2$, then $|q_c(z)| > |z|$ and z is in the stable set of infinity.

b) Use part (a) to show that if $|c| > 2$, then $q_c^n(0)$ tends to ∞.

15.5 LOOKING FOR CANTOR DUST:

It has been suggested that Cantor sets in the complex plane be called Cantor dust since, like a cloud of dust, no two points are connected yet there is another point close to every point in the cloud. In this exercise, we show that the filled Julia set of the function $q_c(z) = z^2 + c$ forms such a Cantor cloud when $|c| > 2$.

Let $|c| > 2$ and define $q(z) = z^2 + c$. Let D be the circle with radius $|c|$ and center at the origin.

a) Use Exercise 15.4a to show that the filled Julia set is contained inside D. In fact, K_c lies inside $q^{-n}(D)$ for all n.

b) Show that $q^{-1}(D)$ is the figure eight with crossing at the origin shown in Figure 15.3.

c) Prove that $q^{-2}(D)$ must be inside $q^{-1}(D)$ and that $q^{-2}(D)$ consists of two figure eights as demonstrated in Figure 15.3.

d) Prove that $q^{-n}(D)$ consists of 2^n figure eights situated as in Figure 15.3. Further, show that as n goes to infinity the diameter of the figure eights in $q^{-n}(D)$ converges to 0.

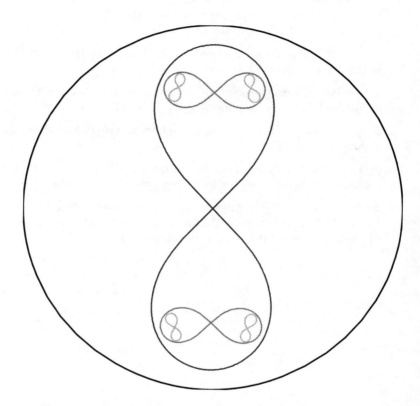

FIGURE 15.3. The inverse images of the disk with radius 2.5 under the map $q_{2.5}$. The filled Julia set is the intersection of all the inverse images of the disk. Successive inverse images are drawn lighter. The nth inverse image consists of 2^n figure eights with one in each of the lobes of the 2^{n-1} figure eights in the $(n-1)$st inverse image.

Appendix
Mathematica Algorithms

Most of the computations for this text were done by *Mathematica* on a NeXT computer. It is easy to use *Mathematica* for computing tables of values of $f^n(x)$, graphing functions, and graphical analysis. It is also easy to use *Mathematica* to generate bifurcation diagrams, the Mandelbrot set, or Julia sets, but it is harder to do so in a reasonable amount of time. The *Mathematica* code included here is fairly efficient and works reasonably well as long as one doesn't try to use too fine a grid. Anyone interested in using *Mathematica* to create a large number of images of bifurcation diagrams, the Mandelbrot set, or Julia sets should seriously consider learning how to use MathLink for the implementation of *Mathematica* being used. Using MathLink as described by W. Shaw and J. Tigg in their text *Applied Mathematica* can decrease the necessary computational time by at least an order of magnitude; their text is listed in the references.

Finally, we note that the author has not stopped working on improving his *Mathematica* code. To get copies of his latest code, contact the author directly by e-mail at rholmgre@alleg.edu or surface mail at Department of Mathematics, Allegheny College, Meadville, PA 16335. Also, you can look for the code in MathSource, a world wide web site at www.wri.com. This site is maintained by Wolfram Research, the makers of *Mathematica* and the author expects to archive his code there.

A.1. Iterating Functions

Finding the Value of a Point Under Iteration. The easiest way to iterate functions with *Mathematica* is to use the Nest command. The format is

   Nest[function, start value, number of iterations].

In the following example, we define $f(x) = x^3$ and then iterate f three times starting at 1.2:

```
Clear[f,x];
f[x_] := x^3;
Nest[f,1.2,3]
```

The output is 137.371.

NestList[...] will show the value of the first n iterates. The format is similar to that of Nest. The following commands compute the first three iterates of 1.2 under $f(x) = x^3$:

```
Clear[f,x];
f[x_] := x^3;
NestList[f,1.2,3]
```

The output is $\{1.2, 1.728, 5.15978, 137.371\}$.

Tables of Iterates. We use the next set of commands to generate a table of iterates. In the first two lines we define the function. Next we initialize variables for the starting value, the first iterate printed, and the last iterate printed. The last block of code is a short *Mathematica* program that does the actual computation. In this example, we are using a While[...] loop to tell *Mathematica* to iterate the function $h(x) = 3.5x(1-x)$ beginning at .1 and to print the values of the 10th through 20th iterates. Note the use of the command N[...] to ensure that we start at a numerical value. If the initial value is expressed as a rational number, for example, 2/3, *Mathematica* will use arbitrary precision arithmetic and this computation will take a *very* long time.

```
Clear[h,x,i,y];
h[x_] := 3.5x (1 - x);

StartingValue = .1;
FirstIteration = 10;
LastIteration = 20;

i=0;
y = N[StartingValue];
While[i <= LastIteration,
    If[ i >= FirstIteration,Print[i,"   ",N[y,8]] ];
```

```
      y = h[y];
      i = i+1 ]
```

The output of this program is as follows:

```
10   0.86414351
11   0.41089828
12   0.84721309
13   0.45305075
14   0.86728519
15   0.40285557
16   0.84197036
17   0.46569696
18   0.87088155
19   0.39356405
20   0.83534986
```

Controlling the Precision of the Computations. For some of the exercises in this text, we need to be able to control the precision of the computations. We use the *Mathematica* command SetPrecision, which writes an expression to n significant digits, adding zeros or rounding as necessary. The format is

SetPrecision[expression,n]

Zeros are added are in binary, so the base ten representation of the new number will not necessarily end in zeros. In the commands that follow, the variable SigDigits controls the amount of rounding:

```
Clear[h,x];
h[x_] := 3.5x(1-x);

StartingValue = .1;
FirstIteration = 10;
LastIteration = 20;
SigDigits = 64;

i=0;
y = SetPrecision[StartingValue,SigDigits];
While[i <= LastIteration,
      If[i >= FirstIteration,Print[i,"   ",N[y,8]]];
      y = SetPrecision[h[y],SigDigits];
      i = i+1]
```

As the function $h(x) = 3.5x(1-x)$ is not sensitive to small changes in initial conditions, the output of this program is the same as that of the preceding program.

Graphing Iterated Functions. For some of the exercises, we need to graph the iterate of a function. For example, in Exercise 8.9, we are asked to graph T^2 where

$$T(x) = \begin{cases} 2x, & \text{for } x \text{ in } [0, \frac{1}{2}] \\ 2 - 2x, & \text{for } x \text{ in } [\frac{1}{2}, 1]. \end{cases}$$

We accomplish this with the following program:

```
Clear[T,x];
T[x_]:= If[x<=.5,2x,2 - 2x];

xmin = 0;
xmax = 1;

NumberOfIterations = 2;
Plot[{Nest[T,x,NumberOfIterations],x},
    {x,xmin,xmax},
    PlotRange->{xmin,xmax},
    AspectRatio->1]
```

Note the use of an If[...] statement to define $T(x)$. The variables xmin and xmax denote the endpoints of the domain. NumberOfIterations = 2 tells the program to compute the graph of T^2. If we replace the 2 with a 3 it will graph T^3. We use Plot and Nest to draw the actual graph. Since we are looking for fixed points of T^2, we draw in the line $y = x$ as well. In the second line of the Plot command, we define the domain of the function to be graphed. PlotRange->{xmin,xmax} tells the computer that the codomain shown should be equal to the domain. Finally, AspectRatio->1 makes the graph square. The output is shown in Figure A.1 on page 213.

A.2. Graphical Analysis

One of the most powerful tools available for investigating the dynamics of functions of the real numbers is graphical analysis. The following program was used to produce Figure A.2a on page 213. We use graphical analysis to analyze the itinerary of .1 under iteration of $h(x) = 3.5x(1 - x)$. The first 20 iterates are shown. Lines enclosed by (* ... *) are comments.

```
Clear[h,x];
h[x_] := 3.5x (1-x)

StartingValue = .1;
FirstIt = 0;
LastIt = 20;
```

```
xmin = 0;
xmax = 1;

    (*  Finding the Points  *)

i = 0;
y = N[StartingValue];

While[i < FirstIt, y = h[y]; i = i + 1];

DataTable = {{y,y},{y,h[y]}};

While[i < LastIt,
     y = h[y];
     AppendTo[DataTable,{y,y}];
     AppendTo[DataTable,{y,h[y]}];
     i = i + 1];

AppendTo[DataTable,{h[y],h[y]}];

    (*  Drawing the Cobweb  *)

Cobweb = ListPlot[DataTable,PlotJoined -> True,
     PlotRange -> {{xmin,xmax},{xmin,xmax}},
     AspectRatio -> 1,PlotStyle->GrayLevel[.3],
     DisplayFunction -> Identity];

    (*  Drawing the graphs of h(x) and y=x  *)

Graph = Plot[{h[x],x},{x,xmin,xmax},
     PlotRange -> {xmin,xmax},AspectRatio -> 1,
     DisplayFunction -> Identity];

    (*  Displaying the result  *)

Show[Cobweb,Graph,DisplayFunction -> $DisplayFunction]
```

To determine the long-term itinerary of .1 under iteration of h, we change the values of FirstIt and LastIt in this program to 100 and 120, respectively. Then the first 100 iterates are calculated, and 20 iterates of the cobweb are drawn beginning with $h^{100}(.1)$. The result is shown in Figure A.2b on page 213; apparently .1 is attracted to a period four orbit.

A.3. Bifurcation Diagrams

The following code was used to generate Figure 12.8; it took approximately five minutes to do so. A similar block of code took more than three hours to generate Figure 12.10.[1] The difference in times is a result of the number of parameter values at which the orbit approached a fixed point. (If $y = h(y)$, the value of i is set to LastIt and the While[...] loop ends.) In generating Figure 12.10, it may have been more efficient to include a test that would determine whether or not a point was already included in the data table before appending it. In general, it is wise to experiment with a small number of parameter values before trying to make a detailed figure. To increase speed, we compiled the function f in the second line. Further gains in processing time could be attained by compiling all or part of the While[...] loops.

```
(*  We define f and h,
        f is compiled for increased speed   *)

f = Compile[{x,c}, (2 x^3 - 1)/(3x^2 + c)];
h[x_] := f[x,p];

        (*  FirstIt is the first iteration graphed,
            LastIt is the last,
            seed is the starting value,
            steps is the number of parameter values used.  *)

FirstIt = 500;
LastIt = 600;
seed = 0.0;
steps = 500

        (*  pmin and pmax define the bounds of the parameters,
            pstep is the stepsize.  *)

pmin =  -2;
pmax =  -.001;
pstep = N[(pmax - pmin)/steps];

        (*  xmin and xmax define the bounds of the values
                to be plotted.  *)
xmin =  -2;
```

---

[1] The author has recently developed new code for bifurcation diagrams that will generate Figure 12.10 in under six minutes. Contact him for details.

```
xmax = 2;
```

  (* We ensure that our parameter is a numerical value
    and create DataTable to store the output *)

```
p = N[pmin];
DataTable = {};
```

  (* Now we determine which points are in the orbit *)

```
While[p <= pmax,
    y = Nest[h,N[seed],FirstIt];
    i = FirstIt;
    While[i <= LastIt,
        AppendTo[DataTable,{p,y}];
        If[y == h[y], i = LastIt, y = h[y]];
        i = i+1];
    p = p + pstep]
```

  (* Finally we use ListPlot to display the points
    stored in DataTable *)

```
ListPlot[DataTable,
    PlotRange -> {{pmin,pmax},{xmin,xmax}},
    AxesOrigin->{pmin,xmin},
    AspectRatio -> 1,
    PlotStyle->{PointSize[.0025]}]
```

The commands PlotRange, AxesOrigin, and AspectRatio customize the output of ListPlot[...] in the obvious way. The optional command PlotStyle->{PointSize[.0025]} improves the quality of the graphics by specifying that the size of the points plotted should be $\frac{1}{400} = .0025$ of the image width. The default value is $\frac{1}{125} = .008$ of the image width, which is a bit too coarse for many plots. To get a good image, it is often worth playing around with the point size.

A.4. Julia Sets

The following code was used to draw the Julia sets of $q_c(z) = z^2 + c$, which are shown in Figure 15.1. Computational time ranged from 30 to 90 minutes. However, by using a coarser grid, we can create informative figures in a small fraction of that time. It is also worth noting that one has to be careful in choosing the number of iterates to be computed before deciding whether or not the point is in, or at least very close to, the Julia set. For example, the Julia set for $q(z) = z^2 + i$, which is shown in Figure 15.1, is very sparse. If we start iterating q at a point in the Julia set, then rounding will soon cause us to leave the Julia set. Thus, within 20 or so iterations, the computed iterate will be well on its way to infinity, even though the actual iterate is still in the Julia set. Consequently, 12 iterations were used to create the image of K_i shown in Figure 15.1.

In the following program, the first block of code defines the compiled function Julia[x,y, p,m]. Julia determines whether or not the point $x + Iy$ leaves the circle of radius $|p| + 1$ before m iterations of $q(z) = z^2 + p$ have been completed. If so, Julia returns the value 1. If not, Julia returns the value 0. The second block of code plots those coordinates (x, y) for which Julia returns the value 0. In this example, we are searching for the Julia set of $q(z) = z^2 - 1$.

The command DensityPlot[...] in the second block of code applies Julia[...] to all the points on a 400 by 400 grid in a box with corners at $-2-2i$, $2-2i$, $2+2i$, and $-2+2i$ to see whether or not they leave the circle with radius 2 within 100 iterations of $q(z)$. If they do leave, then Julia returns the value 1 and the point is left white. Points that don't leave, and hence are in or close to the Julia set, are shown in black. This example took approximately 84 minutes to run. Note that the time of computation increases with the square of the grid size. Consequently, we would expect this same program to run in about 5 minutes if we used a grid size of 100. Decreasing the number of iterations checked also decreases the time needed. And, of course, as we said in the preamble to the Appendix, readers interested in doing many computations should consider using MathLink to decrease the necessary run time by at least another order of magnitude.

```
Julia =
    Compile[{x,y,{p, _Complex},{m, _Integer}},
    Module[{z = x+I y, n = 0, w = p, k = 0},
            bound = Abs[w] + 1;
            While[n <= m && k == 0,
                    If[Abs[z] > bound , k = 1];
                    z = z^2 + w;
                    n++];
        k]
    ];

DensityPlot[Julia[x,y,-1,100],
            {x,-2,2},{y,-2,2},
            PlotPoints -> 400,
            Mesh -> False]
```

There are other methods for computing images of Julia sets. One utilizes the fact that an equivalent definition of a Julia set for the parameter c is the closure of the set of repelling periodic points of $q_c(z) = z^2 + c$. Another uses the fact that for any positive ϵ, we know the circle with radius $1 + |c| + \epsilon$ is not in the filled Julia set K_c. Successive inverse images of this circle are computed. As these inverse images must converge to K_c as the number of iterations goes to infinity, we can often get a good picture quickly. Both of these methods are discussed in the book by Peitgen, Jürgens, and Saupe, which is listed in the references.

A.5. The Mandelbrot Set

The following *Mathematica* program was used to create Figure 15.2. The fundamental design of the program is much the same as that of the program for Julia sets. This program took about 42 minutes to run. As with the Julia set program, the necessary run time increases with the square of the grid size. Run time can also be decreased by decreasing the number of iterations used.

```
Clear[x,y,m,c,z,n,k,r];

Mandel =
    Compile[{x,y,{m, _Integer}},
        Module[{c = x+I y, z = x+I y, n = 0, k = 0,
                    r=Sqrt[x^2 + y^2]},
            If[Abs[c + 1] <= .25,n=m+1];
            If[r == 0, n = m+1,
                If[r <= Abs[(.5/r) c  - (.25/r^2) c^2],n=m+1]];
            While[n <= m && k == 0,
                If[Abs[z] > 2, k=1, z = z^2 + c ];
                n++];
        k]
    ];

DensityPlot[Mandel[x,y,100],
            {x,-2,2},{y,-2,2},
            PlotPoints -> 400,
            Mesh -> False]
```

An alternative method for computing the Mandelbrot set is discussed in the book by Peitgen, Jürgens, and Saupe, which is listed in the references. It uses an inverse iteration scheme similar in concept to the one described earlier for creating Julia sets.

A.6. Stable Sets of Newton's Method

The following program was used to create Figure 14.8. Run time was approximately 4.5 hours. Obviously, this code needs to be optimized further. We refer the reader to the text by Shaw and Tigg for suggestions as to how this might be done. Of course, it was necessary to find the roots of the function before we could write the program, since the three If statements test how close the iterate is to each of the roots.

```
Clear[z,m,w,n,k];
Newton =
    Compile[{{z, _Complex},{m, _Integer}},
        Module[{w = z, n = 0, k = 4},
            While[n <= m && k == 4 && w != 0.0 + 0.0 I,

(*  In the next three lines, we determine how close
the current iterate is to one of the fixed points.
If we are within the desired tolerance, then this
function  returns the assigned value of k.  *)

                    If[Abs[w - 1.0 + 0. I]<=.25,k = 1];
                    If[Abs[w-(-.5+.866I)]<=.25,k = 2];
                    If[Abs[w-(-.5-.866I)]<=.25,k = 3];

(*  The code for Newton's method is added in
the next line.  *)

                    w = (2 w^3 + 1)/(3 w^2);
                    n++];
                k]
        ];

DensityPlot[Newton[ x + I y,50],
{x,-2,2},{y,-2,2},
PlotPoints -> 400,
Mesh -> False]
```

The boundaries of the stable sets of the fixed points of Newton's function for a cubic polynomial are the closure of the set of repelling periodic points. By exploiting this fact, one can write algorithms that compute these boundaries efficiently. We refer interested programmers to the book by Peitgen, Jürgens, and Saupe for suggestions as to how this might be done.

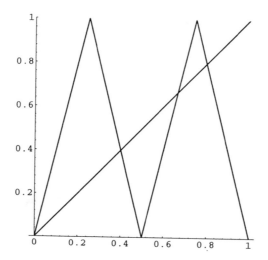

FIGURE A.1. The graph of T^2 and $y = x$ as generated by the program described on page 206.

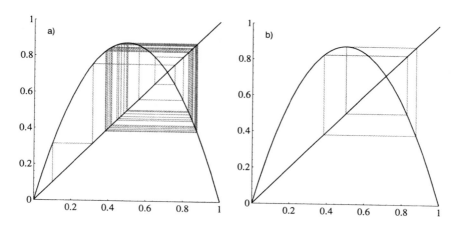

FIGURE A.2. Graphical analysis of the itinerary of .1 under iteration of $h(x) = 3.5x(1-x)$.

References

The items in this bibliography are loosely arranged by topic. In general, only works with which I have had substantial experience are included; this list is by no means exhaustive. My personal reactions to the texts are also included. In addition to sources of information on dynamical systems, I have included references on other mathematical topics mentioned in the text, general interest titles, and computing resources.

Dynamical Systems. New books and articles on dynamical systems appear regularly. Good books on the topic include the following:

DEVANEY, R. *An Introduction to Chaotic Dynamical Systems, 2nd edition*, Addison-Wesley, 1989.

> My text arose out of the desire to provide my students the mathematical background necessary to understand this text by Devaney. His book is good but demands sophistication on the part of the reader. In general, the notation and terminology of my book are consistent with his and the reader should find his book to be a reasonable continuation of the ideas contained in mine.

DEVANEY, R. *A First Course in Chaotic Dynamical Systems: Theory and Experiment*, Addison-Wesley, 1992.

> This text by Devaney covers roughly the same ground as mine but is written from a different point of view. Individuals interested in writ-

ing programs to do computations in dynamics may wish to consult Appendix B of Devaney's book, which contains *TrueBASIC* programs that will do all of the computations necessary to complete the exercises in my text as well as his.

HIRSCH, M. W. AND SMALE, S. *Differential Equations, Dynamical Systems, and Linear Algebra*, Academic Press, 1974

Continuous dynamical systems arise out of the qualitative study of differential equations. This text is an excellent treatment of the qualitative analysis of differential equations and introduces concepts in continuous dynamical systems that are analogous to the concepts we have seen in my book. Hirsch and Smale's book does require sophistication on the part of the reader, but it should be accessible to an advanced undergraduate and is worth the effort. The intertwining of dynamical systems theory and linear algebra with the theory of differential equations motivates their study. Recommended reading for any student interested in pursuing the study of dynamical systems.

HUBBARD, J. AND WEST, B. *Differential Equations: A Dynamical Systems Approach*, Springer-Verlag, 1991

This is a three-volume text that explores the qualitative behavior of differential equations. The fifth chapter (in volume 1) does a good job of introducing numerical methods of solving differential equations as discrete dynamical systems. Limitations of computer implementations of numerical approximations to solutions of differential equations are discussed in Chapter 3. There is a proof of Theorem 12.15 from my text in Section 5.3.

MILNOR, J. *Dynamics in one complex variable: Introductory Lectures*, Stony Brook Institute for Mathematical Sciences Preprint #1990/5

I haven't seen these notes, but I understand that they are a fairly complete set of notes suitable for a first-year graduate-level course on complex dynamics.

PEITGEN, H. AND SAUPE, D., EDS. *The Science of Fractal Images*, Springer-Verlag, 1988

This book should be informative and interesting to anyone who finds my text accessible. It contains numerous algorithms for computing fractal images. In particular, there are several different algorithms for computing Julia sets and the Mandelbrot set.

PEITGEN, H., JÜRGENS, H., AND SAUPE, D. *Chaos and Fractals: New Frontiers in Science*, Springer-Verlag, 1992

> An encyclopedic, though often heuristic, account of many current topics in chaotic dynamics and fractals. I haven't had a lot of time to look at this book, but I have enjoyed those portions that I have read. The last five chapters extend the material covered in my text. In particular, the chapters on Julia sets and the Mandelbrot set present many of the recently discovered properties of these sets. They also include good, efficient programs in BASIC for computing images of these sets. Recommended reading for individuals interested in learning more about fractals, Julia sets, and the Mandelbrot set. The graphics are beautiful, and it is a fun book to peruse.

ROBINSON, C. *Dynamical Systems: Stability, Symbolic Dynamics, and Chaos*, CRC Press, 1995

> An extensive account of many current topics in chaotic dynamics and fractals written at the graduate level. I recently purchased my copy of this text, and I have found it to be very useful. In it, many well-known facts and their proofs are collected together for the first time. In addition the exhaustive references provide a point of departure for anyone wishing to delve more deeply into the subject.

The College Mathematics Journal, Vol. 22, No. 1, January 1991

> This volume of *CMJ* was dedicated to discrete dynamical systems and contains several good articles. Software and computer exercises using ideas from dynamics are also discussed.

General Interest Books on Dynamics.

Gleick, J. *Chaos: Making a New Science*, Viking, 1987

> This book has had a wide readership and students in my courses on dynamics have found it to be good companion reading for a course in dynamics. I prefer the book by Stewart, which is listed next, but there is room for debate. The review of Gleick's book by John Franks and the subsequent responses in *The Mathematical Intelligencer* are worth reading. They help put Gleick's book in perspective and provoke thought about the nature of mathematics. The review and a response are in Vol. 11, No. 1 (winter 1989), pages 65–71. Discussion continues in Vol. 11, No. 3 (summer 1989), pages 3–13.

STEWART, I. *Does God Play Dice?: The Mathematics of Chaos*, Blackwell, 1989

> This is a very good and readable discussion of the implications of modern dynamical systems and chaos theory on the development of science. It is highly recommended as companion reading for a course in dynamical systems or differential equations.

Topics in Mathematics.

BELDING, D., AND MITCHELL, K. *Foundations of Analysis*, Prentice-Hall, 1991

RUDIN, W. *Principles of Mathematical Analysis, 3rd edition*, McGraw-Hill, 1976

> The text by Rudin is an excellent comprehensive introduction to mathematical analysis in $\mathbb{R}^n$ and demands a relatively high degree of mathematical sophistication on the part of the reader. It is definitely on my top ten list of favorite mathematics textbooks. The text by Belding and Mitchell is not as comprehensive, but it is a good, reader-friendly introduction to real analysis. Student consensus is that it is a great book for the first-time student of analysis.

BAK, J., AND NEWMAN, D. J. *Complex Analysis*, Springer-Verlag, 1982

CHURCHILL, R. V., AND BROWN, J. W. *Complex Variables and Applications, 5th edition*, McGraw-Hill, 1990

> The complex analysis texts by Bak and Newman and by Churchill and Brown are both good introductions to complex analysis and are accessible to advanced undergraduates.

WILLARD, S. *General Topology*, Addison-Wesley, 1970

> Willard's textbook on topology is intended for beginning graduate students. Most of the topological ideas covered in my book are adequately covered in the analysis books already mentioned, but individuals interested in all of the details will find Willard's text useful. Personally, I think general topology is fascinating.

OSTROWSKII, A. M. *Solutions of Equations and Systems of Equations, 2nd edition*, Academic Press, 1966

> Ostrowskii's text is an extensive treatment of numerical methods for solving equations. It is included here since it contains a proof of my Theorem 12.15.

Computer Programs and Algorithms. Most of the exercises in this text can be done using *Mathematica*. *Mathematica* is a computer algebra system developed by Wolfram Research of Champaign, Illinois. The programs and commands necessary to use *Mathematica* successfully with this textbook are listed in the Appendix. Helpful books for users of *Mathematica* include the following:

WOLFRAM, S. *Mathematica: A System for Doing Mathematics by Computer, 2nd edition*, Addison-Wesley, 1991

This is the text I turn to first when I have questions about using *Mathematica*.

SHAW, W. T., AND TIGG, J. *Applied Mathematica: Getting Started, Getting It Done*, Addison-Wesley, 1994

This text has helped me dramatically improve my understanding and use of *Mathematica*. Individuals interested in creating images of the Mandelbrot set will want to read Chapter 19, which discusses how one can create good images of the Mandelbrot set in a relatively short period of time. The ideas used also apply to creating images of Julia sets and bifurcation diagrams.

If one doesn't want to use *Mathematica*, there are a multitude of other options one can pursue. Probably the easiest choice for those with some computer experience is to write their own programs. None of the programs for doing the dynamics experiments mentioned in this text need be more than 20 or 30 lines long. Many will have less than 10 lines. To write the programs, one just needs to be able to input initial data to a program, construct loops to do the iteration, and output graphics information to the screen. Several of the books listed in the section on dynamics include computer programs. *TrueBASIC* programs that can easily be converted into other languages are included in Appendix B of the text by Devaney. Numerous programs are included in the book edited by Peitgen and Saupe. Efficient programs in *BASIC* for computing Mandelbrot and Julia sets are found in the book by Peitgen, Jürgens, and Saupe.

For those who want a canned package, the alternatives are almost endless. There are probably hundreds of public domain packages that can do the job. Check local bulletin boards or use your favorite web browser and search the Internet.

Finally, I include the following reference for a workbook and software intended to serve as a supplement to Devaney's text by the same name (listed earlier). While I have never seen it, I believe it contains programs that can complete all the computations necessary for completing the exercises in this text. It requires a Macintosh computer running System 6.0.5 or higher. To get adequate speed, one should have a Mac II and a math coprocessor. For the exact system requirements, please contact Addison-Wesley.

GEORGES, J., JOHNSON, D., AND DEVANEY, R. L. *A First Course in Chaotic Dynamical Systems: Laboratory,* Addison-Wesley, 1992

Index

accumulation point 24, 111
    equivalent conditions for 24, 111
argument of a complex number 169
attracting periodic point 54, 176
    super attracting 133, 148
    weakly attracting 54, 190

bifurcation 59, 60
    diagram 62, 66, 96
    period-doubling 66, 96
    pitchfork 62
    saddle-node 61
    transcritical 63
bifurcation diagram 62, 66, 96
bounded set 73
    in the complex plane 193

C^1 functions 51
Cantor, Georg 84
Cantor dust 201
Cantor set 73, 84
    in complex plane 195, 201
Cantor Middle Thirds Set 73, 85
Cayley, Arthur 184
chaotic 80, 117, 184
    conditions for 81, 117
chain rule 55, 173, 188

circle, doubling map on 124, 136, 150, 178
closed set 25, 111
    equivalent conditions for 26, 112
codomain 9
commutative diagram 87
compact set 73
complement 25
complex functions 170
complex numbers 167
    argument of 169
    arithmetic in 167
    exponential form 170
    modulus of 168
    polar form 170
    on Riemann sphere 178
    square roots of 170
complex plane 168
    extended 181
continuous function 12, 15, 23, 111
    of complex numbers 170, 181
    visualizing 13
convergent sequence 23, 110, 181
converges to infinity 23, 180
critical point 96, 195
    role in dynamics 96, 195

222   Index

dense subset 27, 111
   equivalent conditions for 27, 112
derivative of a function 47, 171
Devaney, R. 80
differentiable 47, 171
distance 13, 110
   in symbol space 109
   on the Riemann sphere 184
domain 9
doubling map 124, 136, 150, 178

ϵ-neighborhood 21
Euler's formula 170
Euler's method 154
eventually fixed point 34
eventually periodic point 34
exponential form of a complex number 170
exponential models 4
extended complex plane 181

family of functions 59
   parametrized 59
Fatou, Pierre 95, 195
Feigenbaum, Mitchell 107
Feigenbaum's constant 100, 107
filled Julia set 194, 211
fixed point 31
   eventually fixed 34
   finding fixed points 31, 33, 48, 50
   hyperbolic 52
forward asymptotic 35
fractal 84
function 9
   C^1 51
   codomain 9
   continuous 12, 170, 181
   differentiable 47, 171
   domain 9
   family of 59
   inverse of 16
   invertible 16
   linear 19
   one-to-one 10
   onto 11
   parametrized family of 59
   range 9

graphical analysis 36
   algorithms for 206
grows without bound 23

homeomorphism 16, 89
   linear functions as 19
hyperbolic attracting set 77
hyperbolic fixed point 52
hyperbolic periodic point 53, 176
hyperbolic repelling set 77

image 10
Intermediate Value Theorem 16
inverse image 10
   of an open set is open 112
invertible function 16
   conditions for 16

Julia, Gaston 195
Julia set 194, 211
   algorithms for 199, 209, 211

least upper bound 121
limit 172
limit point 24
linear fractional transformation 182
logistic function 4, 7, 63

Mandelbrot, Benoit 197
Mandelbrot set 167, 197
   algorithms for 199, 211
map or mapping 9
Mean Value Theorem 47
metric 110
   on Riemann sphere 184
Möbius transformation 182
modulus of a complex number 168
Morse sequence 115

neighborhood 21, 28, 110
   in complex plane 168
   in extended complex plane 180
neutral fixed points 52, 53, 54, 56
Nested Interval Theorem 121
Newton, Isaac 127
Newton's function 129, 182
Newton's method 127
   algorithms for 212
   chaotic behavior in 145, 192
   in the complex plane 182
   for cubic polynomials 138, 186, 191
   for polynomials 131, 182
   for quadratic polynomials 133, 185
Newton-Raphson method 127

nonhyperbolic periodic point 53
  importance of 60

onto 11
  equivalent conditions for 11
one-to-one 10
  equivalent conditions for 10, 16
open set 21, 110, 181
  equivalent conditions for 22, 26, 111, 112
  on Riemann sphere 184
orbit 33
  bounded 193

parameter 59
parametrized family of functions 59
perfect set 73
period 33
  prime period 33
period-doubling bifurcation 66, 96
  cascade of 96
periodic cycle 33
periodic orbit 33
periodic point 33
  attracting 54, 176
  eventually 34
  hyperbolic 53, 176
  repelling 54, 176
  weakly attracting 54, 190
  weakly repelling 54, 190
phase portraits 5
pitchfork bifurcation 62
polar form of a complex number 170
preimage 10
  of an open set is open 112
prime period 33

quadratic map 93, 107, 159
  of the complex plane 178, 193

range of a function 9

Raphson, Joseph 127
repelling periodic points 54, 176
  weakly 54, 190
Riemann sphere 178

saddle-node 61
Sarkovskii, A. N. 43
Sarkovskii's ordering 44
Sarkovskii's Theorem 41, 44
  on an interval 46
sensitive dependence on initial conditions 79, 117
  not preserved by topological conjugacy 91
sequence space 109
  distance in 109
shift map 114
Smale, Stephen 84
stable set 35
  finding stable sets 50, 53, 175
  of a set 78
stereographic projection 179
super attracting 133, 148
symbol space 109
  distance in 109

tent map 86, 92
topological conjugacy 87, 89
  of metric spaces 119
topologically transitive 79, 115
  conditions for 115
topology 21
totally disconnected 73, 195
transcritical bifurcation 63
Triangle Inequality 13, 18, 110

weakly attracting periodic point 54, 190
weakly repelling periodic point 54, 190
well-mixed 79

Universitext *(continued)*

Rubel/Colliander: Entire and Meromorphic Functions
Sagan: Space-Filling Curves
Samelson: Notes on Lie Algebras
Schiff: Normal Families
Shapiro: Composition Operators and Classical Function Theory
Simonnet: Measures and Probability
Smith: Power Series From a Computational Point of View
Smoryński: Self-Reference and Modal Logic
Stillwell: Geometry of Surfaces
Stroock: An Introduction to the Theory of Large Deviations
Sunder: An Invitation to von Neumann Algebras
Tondeur: Foliations on Riemannian Manifolds
Zong: Strange Phenomena in Convex and Discrete Geometry